# Study Guide

## Algebra 1
### Explorations and Applications

This Study Guide includes a section for every section in the textbook. Each Study Guide section contains an illustrated list of Terms to Know, worked-out Examples, and practice exercises, including Spiral Review questions. Each Study Guide chapter contains a Chapter Review consisting of a Chapter Check-Up and a Spiral Review of all previous chapters. Answers are provided in a separate Answer Key.

**McDougal Littell**
**A Houghton Mifflin Company**

Evanston, Illinois
Boston   Dallas   Phoenix

Copyright © 1997 by McDougal Littell Inc. All rights reserved.

No part of this work may be reproduced or transmitted in any form or by any means, electronic or mechanical, including photocopying and recording, or by any information storage or retrieval system without the prior written permission of Houghton Mifflin Company unless such copying is expressly permitted by federal copyright law. Address inquiries to School Permissions, Houghton Mifflin Company, 222 Berkeley Street, Boston, MA 02116

ISBN: 0-395-76963-9

123456789 - HS - 99 98 97 96 95

# CONTENTS

|  |  | *Page* |
|---|---|---|
| | Studying Algebra 1 | vi |

**Chapter 1**
| | | |
|---|---|---|
| Section 1.1 | Working with Variables and Data | 1 |
| Section 1.2 | Using the Order of Operations | 5 |
| Section 1.3 | Mean, Median, and Mode | 9 |
| Section 1.4 | Working with Integers | 12 |
| Section 1.5 | Exploring Negative Numbers | 16 |
| Section 1.6 | Exploring Variable Expressions | 20 |
| Section 1.7 | Applying Variable Expressions | 23 |
| Section 1.8 | Organizing Data | 26 |
| | Chapter 1 Review | 30 |

**Chapter 2**
| | | |
|---|---|---|
| Section 2.1 | Solving One-Step Equations | 32 |
| Section 2.2 | Solving Two-Step Equations | 35 |
| Section 2.3 | Applying Functions | 38 |
| Section 2.4 | Coordinate Graphs | 42 |
| Section 2.5 | Representing Functions | 47 |
| Section 2.6 | Using Graphs to Solve Problems | 51 |
| | Chapter 2 Review | 54 |

**Chapter 3**
| | | |
|---|---|---|
| Section 3.1 | Applying Rates | 56 |
| Section 3.2 | Exploring Direct Variation | 60 |
| Section 3.3 | Finding Slope | 64 |
| Section 3.4 | Finding Equations of Lines | 67 |
| Section 3.5 | Writing an Equation of a Line | 70 |
| Section 3.6 | Modeling Linear Data | 72 |
| | Chapter 3 Review | 75 |

**Chapter 4**
| | | |
|---|---|---|
| Section 4.1 | Solving Problems Using Tables and Graphs | 77 |
| Section 4.2 | Using Reciprocals | 81 |
| Section 4.3 | Solving Multi-Step Equations | 84 |
| Section 4.4 | Equations with Fractions or Decimals | 88 |
| Section 4.5 | Writing Inequalities from Graphs | 92 |
| Section 4.6 | Solving Inequalities | 95 |
| | Chapter 4 Review | 98 |

**Study Guide,** ALGEBRA 1: EXPLORATIONS AND APPLICATIONS

| | | Page |
|---|---|---|
| **Chapter 5** | | |
| Section 5.1 | Ratios and Proportions | **100** |
| Section 5.2 | Scale Measurements | **104** |
| Section 5.3 | Working with Similarity | **108** |
| Section 5.4 | Perimeters and Areas of Similar Figures | **112** |
| Section 5.5 | Exploring Probability | **116** |
| Section 5.6 | Geometric Probability | **119** |
| | Chapter 5 Review | **122** |
| **Chapter 6** | | |
| Section 6.1 | The Pythagorean Theorem | **124** |
| Section 6.2 | Irrational Numbers | **129** |
| Section 6.3 | Calculating with Radicals | **133** |
| Section 6.4 | Multiplying Monomials and Binomials | **136** |
| Section 6.5 | Finding Special Products | **140** |
| | Chapter 6 Review | **143** |
| **Chapter 7** | | |
| Section 7.1 | Using Linear Equations in Standard Form | **145** |
| Section 7.2 | Solving Systems of Equations | **149** |
| Section 7.3 | Solving Linear Systems by Adding or Subtracting | **153** |
| Section 7.4 | Solving Linear Systems Using Multiplication | **157** |
| Section 7.5 | Linear Inequalities | **161** |
| Section 7.6 | Systems of Inequalities | **165** |
| | Chapter 7 Review | **168** |
| **Chapter 8** | | |
| Section 8.1 | Nonlinear Relationships | **170** |
| Section 8.2 | Exploring Parabolas | **173** |
| Section 8.3 | Solving Quadratic Equations | **176** |
| Section 8.4 | Applying Quadratic Equations | **179** |
| Section 8.5 | The Quadratic Formula | **182** |
| Section 8.6 | Using the Discriminant | **186** |
| | Chapter 8 Review | **189** |
| **Chapter 9** | | |
| Section 9.1 | Exponents and Powers | **191** |
| Section 9.2 | Exponential Growth | **195** |
| Section 9.3 | Exponential Decay | **198** |
| Section 9.4 | Zero and Negative Exponents | **202** |
| Section 9.5 | Working with Scientific Notation | **205** |
| Section 9.6 | Exploring Powers | **208** |
| Section 9.7 | Working with Powers | **212** |
| | Chapter 9 Review | **215** |

## Chapter 10

| | | Page |
|---|---|---|
| Section 10.1 | Adding and Subtracting Polynomials | **217** |
| Section 10.2 | Multiplying Polynomials | **220** |
| Section 10.3 | Exploring Polynomial Equations | **223** |
| Section 10.4 | Exploring Factoring | **226** |
| Section 10.5 | Applying Factoring | **229** |
| | Chapter 10 Review | **233** |

## Chapter 11

| | | |
|---|---|---|
| Section 11.1 | Exploring Inverse Variation | **235** |
| Section 11.2 | Using Weighted Averages | **240** |
| Section 11.3 | Solving Rational Equations | **243** |
| Section 11.4 | Applying Formulas | **246** |
| Section 11.5 | Working with Rational Expressions | **249** |
| Section 11.6 | Exploring Rational Expressions | **253** |
| | Chapter 11 Review | **256** |

## Chapter 12

| | | |
|---|---|---|
| Section 12.1 | Exploring Algorithms | **258** |
| Section 12.2 | Finding Paths and Trees | **261** |
| Section 12.3 | Voting and Fair Division | **265** |
| Section 12.4 | Permutations | **269** |
| Section 12.5 | Combinations | **272** |
| Section 12.6 | Connecting Probability and Counting | **276** |
| | Chapter 12 Review | **279** |

# Studying Algebra 1

**STUDY TIPS**

Think about how you learn new material in mathematics. Do you learn best
- **by reading?**
- **by listening?**
- **by writing things down?**
- **by explaining to someone else?**
- **or by some combination of these methods?**

Thinking about what works best for you can help you make the best use of your study time and to remember what you have learned.

Students who learn best by *thinking things through* alone often help themselves remember by putting their thoughts in writing. Making a concept map or writing a paragraph that summarizes the material can help you think about how the information is organized.

Students who learn best by *talking things through* can discuss the material with another student or with a study group. By sharing ideas, students can often help each other learn and remember mathematical concepts.

## *Study Tips*

Here is a list of some activities that may help you learn new material, or review material that you learned previously. You should decide whether each of these is a type of activity that has helped you learn things in the past or that might help you learn things in the future.

1. **Reading**
   - Read the "Learn how to . . ." information at the beginning of each section and see if you can predict the types of problems you will be asked to do.
   - Read the section the night *before* it is presented in class. This can help by focusing your attention on the points that you did not understand when you read about them.
   - Reread the section after it is assigned, looking for any concepts that are still unclear.
   - Reread any notes that you have written and see if you remember what you were thinking about when you wrote them.

2. **Reading and Writing**
   - Read and outline the ideas in the section or chapter.
   - Read each worked-out example and think about the concepts involved; then copy the problem and try to solve it without referring to the solution.
   - Read each worked-out example and then write a similar problem to solve on your own.

**3. Writing**

- Make an outline of the section or chapter.
- Make a concept map for each section.
- Redo any problem you found difficult, comparing your new work with your previous work.
- Write a paragraph that summarizes the section or chapter.
- List the key words from the section and write their definitions in your own words.

**4. Explaining**

- Compare your ideas with those of several other students.
- Form a study group to work together and to help each other.
- Discuss and compare your concept maps for each chapter with other students.
- Identify a few exercises from each section in the chapter that illustrate the main objective of the section.
- Discuss the study techniques that have helped you most in the past with other students.

## *Mathematics Journals*

Writing in a journal helps you keep a record of the things you learn about mathematics, as well as the things you learn about *learning* mathematics. When you write a paragraph, you organize your thoughts and record the important ideas that you have learned.

**Here are some suggestions for writing in your journal that will help you study new material or review material that you learned earlier.**

- Write about your favorite topic in mathematics and how you learned this topic.
- Write a description of the technique that helps you most when you learn new concepts. Explain why you think this helps you learn.
- Write a description of the technique that helps you most when you review a chapter. Did you *write* a summary or list of topics, or did you summarize by *talking* to someone else in your class?
- Look back at your work in previous chapters to see if your technique has changed. If it has changed, how has it changed and why do you think it changed? If it has not changed, explain why this technique seems to be working for you.

# Section 1.1 Working with Variables and Data

**GOAL**

**Learn how to . . .**
- use variables and inequalities

**So you can . . .**
- analyze real-world data

### Application

Shoppers choosing from comparable flight simulator computer games would consider cost. The price is the variable.

Game A . . . . . . . . . . . $50.00
Game B . . . . . . . . . . . $70.00
Game C . . . . . . . . . . . $80.00
Game D . . . . . . . . . . . $55.00

| Terms to Know | Example / Illustration |
|---|---|
| **Data (p. 3)**<br>numbers that give you information | The four prices of the flight simulator computer games discussed in the Application above are data. |
| **Variable (p. 3)**<br>a letter that stands for a quantity that varies | The variable $p$ can be used to represent the price of a flight simulator game in the Application above.<br><br>$p$ = price of game |
| **Value of a variable (p. 3)**<br>the number represented by a variable | If the variable $p$ stands for the price of a game in the Application above, then each price is called a value of $p$.<br><br>for Game A, $p$ = $50.00<br>for Game B, $p$ = $70.00<br>for Game C, $p$ = $80.00<br>for Game D, $p$ = $55.00 |

## UNDERSTANDING THE MAIN IDEAS

### Variables and inequalities

Variables are used in expressions as placeholders for numbers. The value of a variable in an expression can be any number that makes sense in the situation.

When you place the symbol $<$ (is less than), $>$ (is greater than), $\leq$ (is less than or equal to), or $\geq$ (is greater than or equal to) between two numbers you have formed an inequality.

**Study Guide,** ALGEBRA 1: EXPLORATIONS AND APPLICATIONS
Copyright © McDougal Littell Inc. All rights reserved.

### Example 1

Let $c$ stand for the number of calories in a 3 oz serving of meat or poultry.

| 3 oz serving | calories |
|---|---|
| beef rib roast | 375 |
| beef round steak | 165 |
| pork roast | 310 |
| pork chop | 339 |
| broiled chicken | 115 |
| roasted turkey | 151 |

For the set of data above:

**a.** What is the smallest value of $c$?

**b.** What is the largest value of $c$?

**c.** Write two inequalities that describe the data.

### ■ Solution ■

**a.** The smallest value of $c$ is 115.

**b.** The largest value of $c$ is 375.

**c.** All the servings contain 115 calories or more: $c \geq 115$.
All the servings contain 375 calories or less: $c \leq 375$.

---

**For each set of data in Exercises 1 and 2:**

**a. Find the smallest value of the variable.**

**b. Find the largest value of the variable.**

**c. Write two inequalities that describe the data.**

**1.** Let $d$ stand for the number of days in a month.

| Days in a Month ||||||||
|---|---|---|---|---|---|---|---|
| January | 31 | May | 31 | September | 30 |
| February | 28 | June | 30 | October | 31 |
| March | 31 | July | 31 | November | 30 |
| April | 30 | August | 31 | December | 31 |

2. Let t stand for the winning time.

| Olympic Gold Medal Winners 200 Meter Freestyle (Women) | | |
|---|---|---|
| 1992 | Nicole Haislett, U.S. | 1:57.90 |
| 1988 | Heike Friedrich, E. Germany | 1:57.65 |
| 1984 | Mary Wayte, U.S. | 1:59.23 |
| 1980 | Barbara Krause, E. Germany | 1:58.33 |
| 1976 | Kornelia Ender, E. Germany | 1:59.26 |

3. Use the data in Exercise 1. For what percentage of the months is $d \leq 30$?

4. **Open-ended Problem** Determine the playing time of at least 5 songs on your favorite CD or cassette tape. Make a data list. Write two inequalities to describe your data.

5. Let $n$ stand for the number of students in your grade with birthdays in a given month. Give a reasonable value for $n$.

## Reading data from graphs

One way to organize data is in a graph. A bar graph is used to compare data.

### Example 2

**Use the bar graph at the right.**

a. What game has the greatest price?

b. What game has the least price?

c. What is the price of Game B?

d. How many games have a price greater than $50?

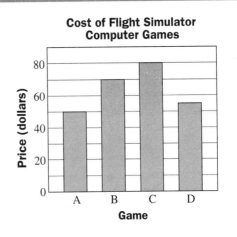

Cost of Flight Simulator Computer Games

### Solution

a. Game C has the greatest price.

b. Game A has the least price.

c. The price of game B is $70.

d. Three games have a price greater than $50.

**Study Guide,** ALGEBRA 1: EXPLORATIONS AND APPLICATIONS
Copyright © McDougal Littell Inc. All rights reserved.

**Use the bar graph at the right.**

6. Let $v$ = the number of electoral votes. Estimate the values of $v$ for the given years.

7. What is the lowest number of electoral votes cast for a winner in one of these presidential elections?

8. How many times did the winner receive more than 400 electoral votes?

9. What is the greatest number of electoral votes cast for a winner in one of these presidential elections?

10. Use the variable $v$ to write two inequalities describing the number of electoral votes cast for the winner in the presidential elections during the years from 1972 to 1992.

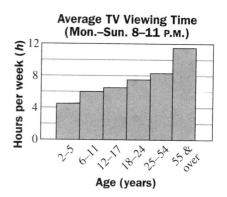

**Use the histogram at the right.**

11. Estimate the average number of hours that 12–17 year olds watch television during the hours of 8–11 P.M., Monday through Sunday.

12. For how many of the given age groups is $h \leq 6$?

13. Which age group spends the most time during this time period watching television?

14. Which age group spends the least time during this time period watching television?

## Spiral Review

**Use the data on gold medal winners in Exercise 2.** *(Toolbox, page 596)*

15. For how many finishes is $t \geq 1$ min 58 s?

16. What is the difference between the fastest time and the slowest time?

**Write each number as a decimal rounded to the nearest hundredth.** *(Toolbox, page 587)*

17. $\dfrac{5}{7}$

18. $\dfrac{11}{9}$

19. $\dfrac{50}{27}$

20. $5\dfrac{2}{3}$

21. $\dfrac{3}{16}$

22. $\dfrac{5}{12}$

# Section 1.2

## Using the Order of Operations

**GOAL**

**Learn how to . . .**
- simplify an expression using the order of operations

**So you can . . .**
- solve real-world problems

### Application

All items at Marieke's favorite store are 50% off. She has chosen three items that cost $19.98, $24.50, and $12.75. Marieke's calculator follows the order of operations. She knows she can divide the regular price by 2 to get the sale price. She uses parentheses around the prices so that her calculator will not divide first. These are her keystrokes:

[(] 19.98 [+] 24.50 [+] 12.75 [)] [÷] 2 [=]

### Terms to Know

| Terms to Know | Example / Illustration |
|---|---|
| **Order of operations (p. 9)**<br>a set of rules people agree to use so that an expression has only one value | To simplify $4 + 3 \cdot 6 \div (3 - 1)$ using the order of operations:<br>1. First do the calculation inside the parentheses.<br>2. Do the multiplication and division in order from left to right.<br>3. Then do the addition.<br>$4 + 3 \cdot 6 \div (3 - 1)$<br>$= 4 + 3 \cdot 6 \div 2$<br>$= 4 + 18 \div 2$<br>$= 4 + 9$<br>$= 13$ |
| **Variable expression (p. 9)**<br>an expression made up of variables, numbers, and operations | $3x + 4$      $n^2 - 6n - 1$<br>$\dfrac{d - c}{5}$ |
| **Evaluate (p. 9)**<br>to replace each variable in an expression with a value and simplify the expression | Evaluating $5r - 2$ for $r = 6$:<br>$5r - 2 = 5(6) - 2$<br>$\phantom{5r - 2} = 30 - 2$<br>$\phantom{5r - 2} = 28$ |

**Study Guide,** ALGEBRA 1: EXPLORATIONS AND APPLICATIONS
Copyright © McDougal Littell Inc. All rights reserved.

**Exponent (p. 10)**
the number in an expression that tells how many times a number or variable is used as a factor

In the expression $x^3$, 3 is the exponent. It tells you to multiply $x$ by itself 3 times.
$x^3 = x \cdot x \cdot x$

## UNDERSTANDING THE MAIN IDEAS

### Order of operations

Since an expression could have several different values depending on the order in which the calculations are done, people have agreed to use a set of rules called the order of operations. The rules are:

1. First do all the work inside parentheses.
2. Then do all multiplications and divisions in order from left to right.
3. Then do all additions and subtractions in order from left to right.

### Example 1

Simplify each expression.

**a.** $2 \cdot 3 + 8 - 4 + 6 \cdot 7$    **b.** $\dfrac{54}{3} - (2 + 6) \cdot 2$

### Solution

**a.** First do the multiplications in order from left to right.

$2 \cdot 3 + 8 - 4 + 6 \cdot 7 = 6 + 8 - 4 + 6 \cdot 7$
$= 6 + 8 - 4 + 42$  ← Now do the additions and subtraction
$= 14 - 4 + 42$         in order from left to right.
$= 10 + 42$
$= 52$

**b.** First do the addition inside the parentheses.

$\dfrac{54}{3} - (2 + 6) \cdot 2 = \dfrac{54}{3} - 8 \cdot 2$  ← Now do the multiplication and division in order from left to right.
$= 18 - 8 \cdot 2$
$= 18 - 16$   ← Finally, do the subtraction.
$= 2$

**Simplify each expression.**

1. $9 \cdot 6 - 4$
2. $35 \div 5 + 6$
3. $6 + 5 \cdot 8$
4. $\dfrac{9 - 5}{2}$
5. $3(6 - 2) + 5$
6. $8 + 8 \div 4$

**7. Open-ended Problem** Simplify the expression 7 • 2 − 1. Then change the operation signs or use parentheses so that the value of the new expression is less than the value of the original.

## Variable expressions

Variable expressions are used in problem-solving situations that involve one or more items that change while the other items remain constant. For example, the speed of a car set on cruise control at 55 mi/h does not change but the number of miles traveled depends on how many hours you drive.

The value of a variable expression depends on the values you substitute for the variables.

### Example 2

Evaluate each variable expression for $r = 6$ and $s = 3$.

a. $r + s$  b. $\dfrac{r - s}{3}$  c. $5(r - s)$

### Solution

a. $r + s = 6 + 3$   ← Substitute 6 for $r$ and 3 for $s$.
$\phantom{r + s} = 9$

b. $\dfrac{r - s}{3} = \dfrac{6 - 3}{3}$   ← Substitute 6 for $r$ and 3 for $s$.
$\phantom{\dfrac{r - s}{3}} = \dfrac{3}{3}$   ← Simplify the numerator.
$\phantom{\dfrac{r - s}{3}} = 1$

c. $5(r - s) = 5(6 - 3)$   ← Substitute 6 for $r$ and 3 for $s$.
$\phantom{5(r - s)} = 5(3)$   ← Do the work inside the parentheses first.
$\phantom{5(r - s)} = 15$

For Exercises 8–13, evaluate each variable expression for $a = 3$ and $b = 9$.

8. $5a + 2b$

9. $7(a + b)$

10. $8b + 6 - (a + b)$

11. $\dfrac{3b - 6}{a}$

12. $\dfrac{2(a + b) - 4}{5}$

13. $(a + b)\left(\dfrac{b - a}{8}\right)$

**14. Open-ended Problem** Many catalog companies base their shipping rate on the purchase price of an item. Write a variable expression for the total cost of an item. Choose several values for the variables and evaluate the expression.

## Exponents

Exponents tell how many times a base is used as a factor. In the expression $4^3$, 4 is the base and 3 is the exponent. To find the value of the expression, 4 is used as a factor 3 times: $4 \cdot 4 \cdot 4 = 64$.

### Example 3

Evaluate $\dfrac{n^2 - (m + n)}{4}$ for $m = 2$ and $n = 6$.

**Solution**

$\dfrac{n^2 - (m + n)}{4} = \dfrac{6^2 - (2 + 6)}{4}$ ← Substitute 2 for $m$ and 6 for $n$.

$= \dfrac{36 - (2 + 6)}{4}$ ← Evaluate the exponent first.

$= \dfrac{36 - 8}{4}$ ← Simplify inside the parentheses.

$= \dfrac{28}{4}$ ← Simplify the numerator.

$= 7$

**Identify the base and exponent in each expression.**

15. $3^6 - 52$
16. $\dfrac{r^3}{2}$
17. $x^8 - 6y$

**For Exercises 18 and 19, evaluate each variable expression for $p = 7$ and $q = 4$.**

18. $p^2 + q^2$
19. $pq + (p^2 - q^2)$

20. **Mathematics Journal** Describe how the value of $x^4$ is different from the value of $4x$.

..................
### Spiral Review

**For each statement, choose a variable and write an inequality.** *(Section 1.1)*

21. More than 700 people attended the graduation ceremonies.
22. Students saw at least 14 different types of birds during a nature walk.
23. The Horticulture Club has planted over 50 species of flowers at the city park.
24. There are less than 250 people signed up for summer activities at the city recreation center this year.

# Section 1.3 Mean, Median, and Mode

**GOAL**

**Learn how to . . .**
- find the mean, the median, and the mode

**So you can . . .**
- analyze real-world data

## Application

The ages of the presidents from 1952–1992 when they took the oath of office were Eisenhower-62 (first term), Kennedy-43, Johnson-55, Nixon-56 (first term), Ford-61, Carter-52, Reagan-69 (first term), Bush-64, and Clinton-46. The average of these ages can be given as the mean or the median. Since no age occurs more than the others, there is no mode.

### Terms to Know / Example / Illustration

| Terms to Know | Example / Illustration |
|---|---|
| **Mean (p. 14)** <br> the number found by dividing the sum of all the data by the number of data items | The mean of the ages in the Application is: <br> $(62 + 43 + 55 + 56 + 61 + 52 + 69 + 64 + 46) \div 9 = 508 \div 9 \approx 56.4$ |
| **Median (p. 14)** <br> the middle number when you put the data in order from smallest to largest (When the number of items is even, the median is the mean of the two middle numbers.) | 43 46 52 55 *56* 61 62 64 69 <br> The median of the presidents' ages is 56. |
| **Mode (p. 14)** <br> The data value that appears most often (There can be no mode or more than one mode) | The mode of the data set 23, 25, 23, 21, 26, and 24 is 23 because it appears twice and all the other data items appear once. |

## UNDERSTANDING THE MAIN IDEAS

The mean, median, and mode are three ways to decscribe the average of a set of data. The mean is often a decimal that needs to be rounded. When you find the median of an even number of data items, there will be two numbers in the middle. In these cases, find the mean of the two numbers. There is no mode if all the items in a data set appear the same number of times.

**Study Guide,** ALGEBRA 1: EXPLORATIONS AND APPLICATIONS
Copyright © McDougal Littell Inc. All rights reserved.

### Example

Use the histogram to find each kind of average.

**a.** the mean
**b.** the median
**c.** the mode

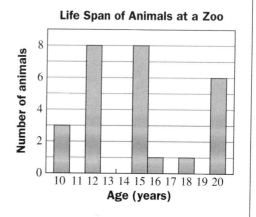

Life Span of Animals at a Zoo

### Solution

**a.** mean = $\dfrac{\text{total number of years}}{\text{number of animals}}$

$= \dfrac{3(10) + 8(12) + 8(15) + 16 + 18 + 6(20)}{27}$

$= \dfrac{30 + 96 + 120 + 16 + 18 + 120}{27}$

$= \dfrac{400}{27}$

$\approx 14.8$

The mean life span of the animals is about 14.8 years.

**b.** To find the median, put the number in order from smallest to largest.

10 10 10 12 12 12 12 12 12 12 12 15 15 15 15 15 15 15 15 15 16 18 20 20 20 20 20 20

   ⎵ 13 items ⎵   ↑ median   ⎵ 13 items ⎵

There are 27 items, so the 14th item is the median: 15 years.

**c.** In a histogram, the mode is represented by the tallest bar. For this data there are two modes, 12 years and 15 years.

---

**Find the mean, the median, and the mode of each set of data.**

**1.** Kilograms Lifted by Gold Medalist Weightlifters in the 1992 Olympics
115, 123, 132, 148, 165, 180, 198, 220, 243

**2.** Years Served by an Individual as Speaker of the House (1899–1989)
4, 8, 8, 6, 6, 2, 2, 1, 4, 7, 2, 4, 2, 6, 9, 6, 10, 2

**3.** Median Weekly Salaries of Non-Union Workers by Age Category (1992)
$272, $450, $410, $484, $496, $447, $354

**Use the histogram at the right.**

4. Find the mean, the median, and the mode of the data in the histogram.

5. **Writing** Does the mean, the median, or the mode best describe the average of the data? Explain.

6. **Open-ended Problem** How would you describe the range of the most typical temperatures found in Milwaukee, WI during the month of July?

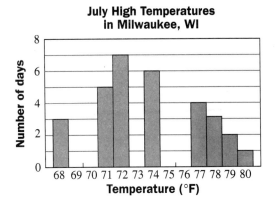

## Spiral Review

**Write each decimal as a fraction in lowest terms. Then write each decimal as a percent.** *(Toolbox, page 587)*

7. 0.75
8. 0.6
9. 0.02
10. 4.00
11. 8
12. 3.25

# Section 1.4 Working with Integers

**GOAL**

**Learn how to...**
- find the absolute value of a number
- add and subtract integers

**So you can...**
- use negative numbers to find differences on number scales, for example

## Application

Meteorologists use integers every time they talk about temperature. "Temperatures are expected to reach 95° today" means +95. "Lows will dip to 10° below zero" means –10.

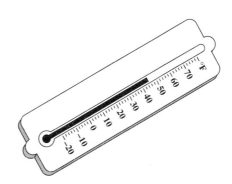

### Terms to Know | Example / Illustration

| Terms to Know | Example / Illustration |
|---|---|
| **Integer (p. 19)**<br>the numbers … –4, –3, –2, –1, 0, 1, 2, 3, 4, … | The numbers on a thermometer are integers. |
| **Absolute value (p. 19)**<br>the distance of a number from zero (The symbol $\|x\|$ means *the absolute value of x*.) | $\|-3\| = 3 \qquad \|8\| = 8$<br>$\|0\| = 0$ |
| **Opposites (p. 20)**<br>two numbers whose sum is zero | The integers 5 and –5 are opposites because 5 + (–5) = 0 and –5 + 5 = 0. |

## UNDERSTANDING THE MAIN IDEAS

### Integers and absolute value

Integers can be positive, negative, or zero. Absolute value tells how far a number is from 0. Since distance is never negative, the absolute value of a number is never negative.

Study Guide, ALGEBRA 1: EXPLORATIONS AND APPLICATIONS
Copyright © McDougal Littell Inc. All rights reserved.

### Example 1

Simplify each expression.

**a.** $3|7-2| + |9|$  **b.** $5|x+3| - |6|$ for $x = 2$

### Solution

**a.** Follow the rules for the order of operations.

$$3|7-2| + |9| = 3|5| + |9|$$
$$= 3 \cdot 5 + 9$$
$$= 15 + 9$$
$$= 24$$

← The absolute value bars act like parentheses, so do the operations inside them first.

**b.** $5|x+3| - |-6| = 5|2+3| - |-6|$ ← Substitute 2 for $x$.
$$= 5|5| - |-6|$$ ← Do the operation inside the absolute value bars first.
$$= 5 \cdot 5 - 6$$
$$= 25 - 6$$
$$= 19$$

**Simplify each expression.**

**1.** $2|9+7| - |-12|$  **2.** $|-16| + |16|$  **3.** $|-4| + 3|6-2|$

**Evaluate each expression for $x = 5$.**

**4.** $|x+2| - 4$  **5.** $2|x-1| + 5$  **6.** $|x+6| - |x+6|$

## Adding and subtracting negatives

Every number has an opposite. The numbers 10 and –10 are opposites because $10 + (-10) = 0$ and $-10 + 10 = 0$.

Subtracting a number is the same as adding its opposite: $16 - 9 = 16 + (-9)$.

### Example 2

Simplify each expression.

**a.** $4 + (-2)$  **b.** $5 - 9$  **c.** $10 - (-3)$

### Solution

Two-color chips can be used to add integers.

Let ● stand for 1 and ○ stand for –1.

● and ○ are called a *zero pair* because their sum is 0.

**a.** 4 + (–2) can be modeled as:

● ● ● ●

○ ○

Now remove the two zero pairs.

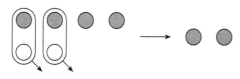

Removing the two zero pairs shows that 4 + (–2) = 2.

**b.** Since subtracting is the same as adding the opposite, 5 – 9 can be rewritten as the addition 5 + (–9).

$$5 + (-9) = 5 + [-5 + (-4)] \quad \leftarrow \text{Think of } -9 \text{ as } -5 + (-4).$$
$$= [5 + (-5)] + (-4)$$
$$= 0 + (-4)$$
$$= -4$$

**c.** Rewrite the subtraction 10 – (–3) as the addition 10 + 3.

$$10 - (-3) = 10 + 3$$
$$= 13$$

**Simplify each expression.**

**7.** 3 – 12      **8.** 7 – 11      **9.** 15 + (–6)

**10.** 7 + (–2)      **11.** 9 – (–12)      **12.** 6 – (–3)

**13.** $5 - |6|$      **14.** $2|1-6| + 3$      **15.** $|8-9| + 2|4| - 7$

**For Exercises 16–18, evaluate each variable expression for $x = 6$, $y = 2$, and $z = 5$.**

**16.** $x + (-y)$      **17.** $3y - 4z + x$      **18.** $|x - y - z|$

**19. Oceanography** A scuba diver is currently at a depth of 140 ft. She ascends at a rate of 60 ft/min. She must make a decompression stop at a depth of 20 ft. How long will it take her to reach that stop?

## Spiral Review

**Find the perimeter and area of each figure.** (*Toolbox, page 593*)

20.
square

21.
rectangle

22.
circle

23.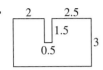

24.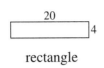
rectangle

25.
triangle

# Section 1.5 Exploring Negative Numbers

**GOAL**

**Learn how to...**
- multiply and divide integers
- use properties of addition and multiplication

**So you can...**
- solve problems involving negative numbers

## Application

Different diving methods allow divers to reach various depths. A breath-hold diver can stay at depths of 40 ft for 1 min while a scuba diver can spend 5 min at 160 ft.

Ocean depths are represented by integers. The depth reached by the scuba diver is 4 times that reached by the breath-hold diver.

$$-160 \div (-40) = 4$$

## UNDERSTANDING THE MAIN IDEAS

### Multiplying and dividing integers

When you multiply (or divide) two integers with the same sign, the product (or quotient) is positive. When you multiply (or divide) two integers with different signs, the product (or quotient) is negative.

### Example 1

Evaluate each expression for $r = -2$.

a. $-3r + 6$  b. $5r - 4$  c. $\dfrac{-9r}{-6}$  d. $\dfrac{6r}{-3}$

### ■ Solution ■

**a.** Substitute $-2$ for $r$. Then follow the order of operations.

$-3r + 6 = (-3)(-2) + 6$  ← Note: $-3r$ means $(-3)(-2)$, not $-3 - 2$.

$\quad = 6 + 6$  ← $-3$ and $-2$ have the same sign, so the product is positive.

$\quad = 12$

**b.** $5r - 4 = (5)(-2) - 4$  ← Substitute $-2$ for $r$.

$\quad = -10 - 4$  ← $5$ and $-2$ have different signs, so the product is negative.

$\quad = -14$

*(Solution continued on next page.)*

### ■ Solution ■ (continued)

c. $\dfrac{-9r}{-6} = \dfrac{-9(-2)}{-6}$  ← Substitute –2 for r.

$= \dfrac{18}{-6}$  ← Simplify the numerator first: –9 and –2 have the same sign.

$= -3$  ← 18 and –6 have different signs, so the quotient is negative.

d. $\dfrac{6r}{-3} = \dfrac{6(-2)}{-3}$  ← Substitute –2 for r.

$= \dfrac{-12}{-3}$  ← Simplify the numerator first: 6 and –2 have different signs.

$= 4$  ← –12 and –3 have the same sign, so the quotient is positive.

**Simplify each expression.**

1. $(6)(-3)$
2. $-5(-8)$
3. $9(-4)$
4. $\dfrac{32}{-8}$
5. $-63 \div 9$
6. $\dfrac{-14}{-7}$

**For Exercises 7–12, evaluate each expression for $s = -2$.**

7. $3s$
8. $-4s$
9. $2s - 8$
10. $\dfrac{8s}{4}$
11. $-3s - 10$
12. $\dfrac{-5s}{5}$

13. That frozen pizza you had for a snack could have been processed in an air blast freezer or a cryogenic freezer. Airblast freezers reach temperatures of about –40°F. The temperature of a cryogenic freezer is around –280°F. How many times colder is a cryogenic freezer than an airblast freezer?

## *Properties of addition and multiplication*

Using properties can help you to simplify an expression. The commutative property and the associative property are described below.

*Commutative Property*
You can add or multiply numbers in any order.

        Addition:             Multiplication:
        $5 + (-1) = -1 + 5$    $(5)(-1) = (-1)(5)$

*Associative Property*
When you add or multiply three or more numbers, you can group the numbers without changing the result.

        Addition:             Multiplication:
        $(4 + 3) + 6 = 4 + (3 + 6)$    $(4 \cdot 3) \cdot 6 = 4 \cdot (3 \cdot 6)$

### Example 2

**Simplify each expression.**

**a.** (0.12)(–3) • 100   **b.** 3 + 5 – 10 – 5

### ■ Solution ■

**a.** It may be easier to multiply 0.12 by 100 than 0.12 by –3.

(0.12 • –3) • 100 = (–3 • 0.12) • 100   ← Use the commutative property.
= 3 • (0.12 • 100)   ← Use the associative property.
= –3 • 12
= –36

**b.** Remember that subtracting is the same as adding the opposite.

3 + 5 – 10 – 5 = 3 + 5 + (–10) + (–5)   ← Rewrite the subtractions as additions.
= 3 + (–10) + 5 + (–5)   ← Use the commutative property.
= [3 + (–10)] + [5 + (–5)]   ← Group the numbers so they are easier to add.
= –7 + 0   ← 5 and –5 are opposites.
= –7

**Simplify each expression.**

**14.** 3 – 4 + (–3) + 11   **15.** $\left(\frac{3}{4} \cdot 9\right)(-12)$   **16.** $\frac{(-4 - 11)(15)}{5}$

**For Exercises 17–20, tell whether the expressions are *equal* or *not equal*. If they are equal, tell whether they demonstrate the *associative property*, the *commutative property*, or *both*.**

**17.** –3 + 6 + 5 = 5 + (–3) + 6   **18.** (24 ÷ 6) ÷ 2 = 24 ÷ (6 ÷ 2)

**19.** (8 – 5) – 2 = 8 – (2 – 5)   **20.** (2 • 3) • 8 = 2 • (3 • 8)

**21. Open-ended Problem** Write an expression that can be simplified in at least two ways. Describe two methods using the commutative or associative properties.

## Spiral Review

**For each pair of numbers, find the greatest common factor and the least common multiple.** *(Toolbox, page 584)*

**22.** 3 and 18  **23.** 12 and 32  **24.** 16 and 48

**Write each fraction as a decimal and a percent.** *(Toolbox, page 587)*

**25.** $\frac{4}{9}$  **26.** $\frac{1}{6}$  **27.** $\frac{32}{25}$

# Section 1.6

## Exploring Variable Expressions

**GOAL**

**Learn how to...**
- write variable expressions

**So you can...**
- solve real-world problems about nutrition, for example

### Application

Gulliver's Bike Shop charges a handling fee plus an hourly rate of $25 for repairs. A variable expression can be used to describe the repair costs.

**GULLIVER'S Bike Shop**

Repair Charges*
$25 per hour

*There is a $10 handling fee on all bike repairs.

### UNDERSTANDING THE MAIN IDEAS

Patterns can help you write variable expressions and solve problems. You can identify a pattern by making an organized list.

#### Example

a. Write a variable expression for the cost of repairs at Gulliver's Bike Shop in the Application above.

b. Tickets for the Shakespeare Festival are $10.00 for adults and $6.00 for students. Write a variable expression for the price of tickets.

#### Solution

a. Make an organized list and look for a pattern. Use repair hours that are reasonable.

| Number of hours | Total cost |
|---|---|
| 1 | 25(1) + 10 = $35 |
| 2 | 25(2) + 10 = $60 |
| 3 | 25(3) + 10 = $85 |
| 4 | 25(4) + 10 = $110 |
| ... | ... |

Use the pattern to write a variable expression. If it takes $h$ hours to repair a bicycle, then the cost of the repairs at Gulliver's Bike Shop is $25h + 10$.

*(Solution continued on next page.)*

Study Guide, ALGEBRA 1: EXPLORATIONS AND APPLICATIONS
Copyright © McDougal Littell Inc. All rights reserved.

### ■ Solution ■ (continued)

**b.** Make an organized list and look for a pattern. Think of some likely combinations of people who would go to the festival. Then find the cost of the tickets for each group.

| Number of adults | Number of students | Cost of tickets |
|---|---|---|
| 1 | 0 | 10(1) + 6(0) = $10 |
| 1 | 1 | 10(1) + 6(1) = $16 |
| 1 | 2 | 10(1) + 6(2) = $22 |
| 2 | 0 | 10(2) + 6(0) = $20 |
| 2 | 1 | 10(2) + 6(1) = $26 |
| 2 | 2 | 10(2) + 6(2) = $32 |

Use the pattern to write a variable expression. Letting $a$ = the number of adults and $s$ = the number of students, the total cost of $a$ adult tickets and $s$ student tickets for the festival is $10a + 6s$.

---

1. Use the movie theatre ticket sign shown at the right. Write a variable expression for the total cost of tickets for $a$ adults, $c$ children, and $s$ senior citizens.

**TICKET PRICES**
(All Shows)
Adults ................ $6.50
Children
(under 13) ............ $4.50
Senior Citizens
(over 55) ............ $5.00

2. Write a variable expression for the final score in **(a)** a football game and **(b)** a basketball game.

3. The Pep Club is selling T-shirts for $12.95 each. There is a $2.00 extra charge for each XXL size T-shirt ordered.

   **a.** Write a variable expression for the total cost of an order that includes both regular and XXL T-shirts.

   **b.** The club earns $3.50 profit on each T-shirt they sell. Write a variable expression for their profit.

4. Saburo and several of his friends bought ice cream cones at the Super Soft-Serve. Write a variable expression for the total cost of their ice cream cones.

Small $.89   Medium $.99   Large $1.19

5. Miranda has collected a total of $50 per mile in pledges for the Union River Romp, a 10-mile walk-a-thon. Her school will receive 50% of the amount she raises. Write a variable expression for the amount Miranda's school will receive.

6. The school cafeteria is serving chicken nuggets for lunch. They will serve 6 to each student who has lunch. Write a variable expression for the minimum number of chicken nuggets the kitchen staff will need to prepare.

7. Keifer earns $5.75 an hour at his job plus $8.65 for every hour of overtime. He pays his employer $3.50 each week for his uniforms. Write a variable expression for his weekly pay.

**Open-ended Problem** Describe a situation that each variable expression can represent.

**8.** $h + 7.50$  **9.** $6t$  **10.** $1.5c$

**11.** $\dfrac{d}{11}$  **12.** $4s + 5$  **13.** $.75s + .99m + 1.25l$

## Spiral Review

**Evaluate each expression for $r = 5$.** *(Section 1.2)*

**14.** $0.75r$  **15.** $4r + 7$  **16.** $3(r - 2)$  **17.** $4(12 - 3r)$

**18.** $6(r - 9)$  **19.** $r(-12 + 6)$  **20.** $\dfrac{3r - 7}{4}$  **21.** $20 - (-5r)$

# Section 1.7 Applying Variable Expressions

**GOAL**

**Learn how to...**
- simplify variable expressions
- use the distributive property

**So you can...**
- apply variable expressions more easily in problem solving situations

## Application

Students in a Life Skills class are making a large 9-cell playing board from canvas. They plan to use individual canvas pieces for each cell and trim them with fabric. To be sure they have enough trim they will purchase 12 extra inches per cell. Let $s$ stand for the length of a cell side in inches. The variable expression $9(4s + 12)$ can be used to determine how many inches of trim they will purchase.

### Terms to Know / Example / Illustration

| Terms to Know | Example / Illustration |
|---|---|
| **Equivalent expressions (p. 35)** <br> expressions that are equal no matter what value you substitute for the variable(s) | The expressions $2(x - 3)$ and $2x - 6$ are equivalent. |
| **Terms (p. 36)** <br> the parts of a variable expression that are added together | The four terms in the expression $4x^2 - 2x + (6x - 7)$ are $4x^2$, $-2x$, $6x$, and $-7$. |
| **Like terms (p. 36)** <br> terms with identical variable parts | $4x^2 - 2x + (6x - 7)$ <br>    ↑  ↑ <br>   like terms |
| **Coefficient (p. 36)** <br> the numerical part of a term | The coefficient of the term $4x^2$ is 4. |
| **Simplest form (p. 36)** <br> the form of an expression in which there are no parentheses and all like terms are combined | The simplest form of the expression $4x^2 - 2x + (6x - 7)$ is $4x^2 + 4x - 7$. |

# UNDERSTANDING THE MAIN IDEAS

## Equivalent expressions

In order to simplify an expression, it is often necessary to write an equivalent expression. The distributive property, $a(b + c) = ab + ac$, can be used to write equivalent expressions.

### Example 1

Simplify each variable expression by first using the distributive property to write an equivalent expression.

**a.** $-(2t + 1)$      **b.** $\frac{1}{5}(t - 20)$

### Solution

**a.** When there is a minus sign in front of a quantity in parentheses, think of it as $-1$ times the quantity.

$-(2t + 1) = -1(2t + 1)$    ← Use the multiplicative property of $-1$: $a(-1) = -a$ and $(-1)a = -a$.

$= (-1)(2t) + (-1)(1)$    ← Use the distributive property.

$= -2t + (-1)$, or $-2t - 1$

**b.** $\frac{1}{5}(t - 20) = \frac{1}{5}[t + (-20)]$    ← Subtracting 20 is the same as adding $-20$.

$= \left(\frac{1}{5}\right)(t) + \left(\frac{1}{5}\right)(-20)$    ← Use the distributive property.

$= \frac{1}{5}t + (-4)$, or $\frac{t}{5} - 4$

**Simplify each variable expression.**

1. $2(x + 5)$
2. $-(3r - 8)$
3. $4(3s + 7)$
4. $\frac{1}{2}(8t - 12)$
5. $-3(5y + 6)$
6. $0.75(16x - 4)$
7. $9(r + 3)$
8. $\frac{1}{3}(9r + 21)$
9. $-8(2x - 7)$

## Simplifying variable expressions

The numerical part of a term is called the coefficient. To combine like terms, you add their coefficients. An expression is in simplest form when it contains no parentheses and no like terms remain.

### Example 2

**Simplify each variable expression by combining like terms.**

a. $3x + 9 + 2x + 3$
b. $2x + y + 8$
c. $4x + 7 - 2x + 2y + 5$
d. $3x^2 - 4 + 2x - x^2 + 3x - 8$

### Solution

a. $3x + 9 + 2x + 3 = (3x + 2x) + (9 + 3)$ ← Use the associative property to group like terms.

$= 5x + 12$ ← Add the coefficients of $x$.

b. There are no like terms or parentheses in the expression $2x + y + 8$. The expression cannot be simplified.

c. $4x + 7 - 2x + 2y + 5 = (4x - 2x) + 2y + (7 + 5)$ ← Use the associative property to group like terms.

$= 2x + 2y + 12$

d. $3x^2 - 4 + 2x - x^2 + 3x - 8 = (3x^2 - x^2) + (2x + 3x) + [-4 + (-8)]$

---

**For Exercises 10–18, simplify each variable expression.**

10. $13a + 4a$
11. $7n - 6 + 18n$
12. $9r + 27 + 3$
13. $-3x + 2 + 14x + 29$
14. $7k - 3m + 6k + 9m$
15. $9x + 2x^2 - 7x$
16. $n^2 - 2n + 2$
17. $3a - 5 - a + 2$
18. $3z^2 + 4z - 7 + 5z^2$

19. **Mathematics Journal** On Monday, James sold 3 CDs and a poster to his first customer. Then he sold 1 CD and 2 posters to his next customer. Describe the steps you would follow to **(a)** write a variable expression for each sale and **(b)** combine the two expressions to show the total sales.

### Spiral Review

**Evaluate each variable expression for $x = 3$ and $y = -6$.** *(Section 1.5)*

20. $4x - 2$
21. $8y - 3y$
22. $3x - 2y + 7$
23. $3(5x - 2)$
24. $\dfrac{x - 9}{y}$
25. $\dfrac{5(2x + 7y)}{12}$

# Section 1.8 Organizing Data

**GOAL**

**Learn how to . . .**
- organize data using spreadsheets and matrices

**So you can . . .**
- work conveniently with large amounts of data

### Application

Manufacturers spend millions of dollars each year to advertise their products. The matrices below show how some of those advertising dollars were spent in 1991 and 1992. The amounts are in millions of dollars.

$$A = \begin{array}{r} \text{Retail} \\ \text{Auto} \\ \text{Food} \\ \text{Cosmetics} \end{array} \begin{array}{ccc} \text{Print} & \text{TV} & \text{Radio} \\ \begin{bmatrix} 2782 & 1962.7 & 92.1 \\ 1724.8 & 3278.5 & 72.2 \\ 499 & 2904.9 & 41.9 \\ 660.5 & 1559.7 & 8.2 \end{bmatrix} \end{array}$$

$$B = \begin{array}{r} \text{Retail} \\ \text{Auto} \\ \text{Food} \\ \text{Cosmetics} \end{array} \begin{array}{ccc} \text{Print} & \text{TV} & \text{Radio} \\ \begin{bmatrix} 5214.4 & 2085.9 & 71.3 \\ 1977 & 3626.6 & 50.9 \\ 542 & 2850 & 34.4 \\ 754.8 & 1653.2 & 7.0 \end{bmatrix} \end{array}$$

## Terms to Know / Example / Illustration

| Terms to Know | Example / Illustration |
|---|---|
| **Matrix (p. 40)** — a group of numbers arranged in rows and columns (plural: *matrices*) | $\begin{bmatrix} 5 & -6 & 1 & 0 \\ 2 & -1 & -5 & 3 \\ 8 & 0 & 0 & 4 \end{bmatrix}$ |
| **Element (p. 40)** — each number in a matrix | There are 12 elements in the matrix above. |
| **Dimensions (p. 40)** — in a matrix, the number of rows and the number of columns, in that order | The dimensions of the matrix above are $3 \times 4$ (3 rows by 4 columns.) |

## UNDERSTANDING THE MAIN IDEAS

A matrix is an arrangement of elements in rows and columns. The plural of matrix is matrices. The numbers in a matrix are placed inside large brackets. All labels are written outside the brackets. A capital letter can be used to identify a matrix.

### Example 1

Use the two matrices in the Application.

**a.** Give the dimensions of each matrix.

**b.** What element in matrix $B$ is in the same position as the element 2904.9 in matrix $A$?

**c.** Find the combined 1991 and 1992 cost of advertising in each category.

### ■ Solution ■

**a.** Each matrix has 4 rows and 3 columns, so the dimensions are $4 \times 3$.

**b.** The element 2904.9 is in the third row, second column of matrix $A$. The element in the same position in matrix $B$ is 2850.

**c.** Adding matrices $A$ and $B$ will give the combined costs. To add the matrices, add each element in matrix to $B$ to the element in the same position in matrix $A$.

$$\begin{bmatrix} 2782 & 1962.7 & 92.1 \\ 1724.8 & 3278.5 & 72.2 \\ 499 & 2904.9 & 41.9 \\ 660.5 & 1559.7 & 8.2 \end{bmatrix} + \begin{bmatrix} 5214.4 & 2085.9 & 71.3 \\ 1977 & 3626.6 & 50.9 \\ 542 & 2850 & 34.4 \\ 754.8 & 1653.2 & 7.0 \end{bmatrix} = \begin{bmatrix} 7996.4 & 4048.6 & 163.4 \\ 3701.8 & 6905.1 & 123.1 \\ 1041 & 5754.9 & 76.3 \\ 1415.3 & 3212.9 & 15.2 \end{bmatrix}$$

The last matrix gives the cost (in millions of dollars) of advertising in each category. For example, the cosmetics industry spent a total of $3212.9 million for advertising on television during 1991 and 1992.

**Use matrices A, B, and C below.**

$$A = \begin{bmatrix} 27 & 43 & 18 \\ 16 & 55 & 25 \\ 21 & 48 & 29 \end{bmatrix} \qquad B = \begin{bmatrix} 29 & 48 & 15 \\ 22 & 45 & 32 \\ 33 & 41 & 27 \end{bmatrix} \qquad C = \begin{bmatrix} 342 & 557 \\ 288 & 545 \\ 301 & 5754 \end{bmatrix}$$

**1.** Give the dimensions of each matrix.

**2.** Add matrices $A$ and $B$.

**3. Writing** Describe why matrix $C$ cannot be added to matrix $A$.

**4. Open-ended Problem** Write a matrix with the same dimensions as matrix $C$.

## Spreadsheets

A spreadsheet is a computer software application that organizes data in rows and columns, and is often used to perform calculations on the data. When you open a spreadsheet file you will find an open chart for organizing your data in rows and columns. The rows are labeled with numbers. The columns are labeled with letters. The boxes on the spreadsheet are called *cells*. When you highlight a cell, its contents appear in the formula bar.

### Example 2

Use the spreadsheet below.

| | A | B | C | D |
|---|---|---|---|---|
| 1 | | 1993 | 1994 | % Increase |
| 2 | Elementary | 4457 | 9791 | 119.67691 |
| 3 | Jr. High | 2326 | 4261 | 83.190026 |
| 4 | Senior High | 4168 | 6713 | 61.060461 |
| 5 | Total | 11021 | 20943 | 90.028128 |
| 6 | | | | |

(CD-ROMS.SS)

a. What does cell C3 show?

b. What is the formula for Cell D2? What does the cell show?

#### Solution

a. Cell C3 is in column C, row 3. It shows the number of CD-ROMs in Junior High Schools in 1994.

b. The formula is (C2–B2/B2)*100. It shows the percent of increase in the number of CD-ROMs in elementary schools from 1993 to 1994.

---

**For Exercises 5–7, use the spreadsheet on Algebra 1 grades shown below.**

| | A | B | C | D |
|---|---|---|---|---|
| 1 | | 1st Quarter | 2nd Quarter | Semester Avg |
| 2 | Alyssa | 88 | 94 | |
| 3 | Shannon | 75 | 82 | |
| 4 | James | 95 | 86 | |
| 5 | Miko | 85 | 98 | |
| 6 | Sarah | 89 | 88 | |
| 7 | | | | |

(GRADES.SS)

**5.** What does cell B6 show?

**6.** What is the formula for cell D2? What will the cell show?

**7. Open-ended Problem** What other calculations could be made using this spreadsheet?

### Example 3

Refer to the information in Example 2. Use a spreadsheet to find the percent of increase in the number of CD-ROMs in Junior High Schools from 1993 to 1994.

> ■ **Solution** ■
>
> **Step 1** Enter all the data on your spreadsheet.
>
> **Step 2** Click on cell D3.
>
> **Step 3** Enter a formula to tell the computer how to find the percent of increase in each type of school. The formula for percent increase is:
>
> $$\frac{\text{number of CD-ROMs in 1994} - \text{number of CD-ROMs in 1993}}{\text{number of CD-ROM in 1993}} \cdot 100$$
>
> ((C2–B2)/B2)*100  ← The symbol * is used for multiplication
>
> **Step 4** Highlight cells D2 through D4. Then select "Fill Down" under Edit.
>
> The percent of increase in the number of CD-ROMs in Junior High Schools from 1993 to 1994 is about 83%.

Each year the NBA publishes statistics on their top players. For Exercises 8–11, use the 1993–1994 statistics on field goals and field goal attempts given at the right.

| | A | B | C |
|---|---|---|---|
| 1 | | FG | FGA |
| 2 | O'Neal, Orlando | 953 | 1591 |
| 3 | Mutombo, Denver | 365 | 642 |
| 4 | Thorpe, Houston | 449 | 801 |
| 5 | Webber, Golden State | 572 | 1037 |
| 6 | Kemp, Seattle | 533 | 990 |
| 7 | Vaught, L.A. Clippers | 373 | 695 |
| 8 | Ceballos, Phoenix | 425 | 795 |
| 9 | Smits, Indiana | 493 | 923 |
| 10 | D. Davis, Indiana | 308 | 582 |
| 11 | Olajuwon, Houston | 894 | 1694 |
| 12 | Stockton, Utah | 458 | 868 |

8. Write a matrix showing the field goal statistics of the top NBA players for the 1993–1994 season.

9. What are the dimensions of the matrix?

10. What formula would you enter to find Shaquille O'Neal's field goal percentage?

11. Use a spreadsheet to find the field goal percentage for each of the top NBA players for the 1993–1994 season.

## Spiral Review

**Evaluate each expression for** $a = -3$, $b = 0.25$, **and** $c = 4$. *(Section 1.5)*

12. $3a + c$

13. $12b$

14. $2(c - a)$

15. $b - a$

16. $bc$

17. $\dfrac{ac}{-6}$

# Chapter 1 Review

**CHAPTER CHECK-UP**

Complete these exercises for a review of Chapter 1. If you have difficulty with a particular problem, review the indicated section.

For Exercises 1–3, use the histogram at the right. *(Section 1.1)*

1. How many significant hurricanes occurred from 1954–1966?
2. How many significant hurricanes occurred from 1928–1992?
3. Find the mode of the number of significant hurricanes occurring in the 12-year intervals given.

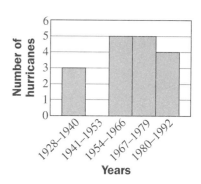

For Exercises 4–6, simplify each expression. *(Section 1.2)*

4. $(8 - 2) \cdot 1 + 0$
5. $8(5 - 1) + 2$
6. $2 \cdot 5 - \dfrac{8}{1 + 3}$

7. **Open-ended Problem** Find the mean, the median, and the mode of the ages of the people living in your home. *(Section 1.3)*

8. **Writing** Which kind of average best describes the data in Exercise 7? *(Section 1.3)*

Evaluate each variable expression for $a = 2$, $b = -3$, and $c = 5$. *(Section 1.4)*

9. $-(a + b)$
10. $|b| + |c|$
11. $4a - |c| - |b|$

Evaluate each expression for $x = -2$. *(Section 1.5)*

12. $-5(x + 6)$
13. $|6x|$
14. $\dfrac{4x - 12}{5}$

For Exercises 15 and 16, refer to the coupon at the right. *(Section 1.6)*

15. Write a variable expression for the cost of each pizza in the promotion.
16. Write a variable expression for the difference between the regular price and the promotion price.

---- Coupon ----
**VILLAGE PIZZA**

Buy 2 pizzas at the regular price and get a third pizza of the same size FREE!
---- Coupon ----

For Exercises 17 and 18, simplify each variable expression. *(Section 1.7)*

17. $4a^2 + 2a + 3 + 4a$
18. $3(a - 2) + 5a + 7$

19. **Open-ended Problem** Use the distributive property to write two equivalent expressions. *(Section 1.7)*

**Use matrices A and B.** *(Section 1.8)*

$$A = \begin{bmatrix} 6 & 8 & 7 \\ 3 & 9 & 2 \\ 4 & 0 & 5 \end{bmatrix} \qquad B = \begin{bmatrix} 4 & 8 & 5 \\ 9 & 7 & 2 \\ 1 & 0 & 6 \end{bmatrix}$$

**20.** Give the dimensions of each matrix.

**21.** Add the matrices.

**Use the spreadsheet of batting records for the 1994 Chicago White Sox.** *(Section 1.8)*

**22.** The batting average of a baseball player is the ratio of hits to the number of at bats. What formula could you enter to find Frank Thomas's batting average?

**23. Writing** Describe the steps you would follow to use a spreadsheet to find batting averages for all the players.

| | A | B | C | D |
|---|---|---|---|---|
| | | CHICAGO.SS | | |
| 1 | player | at bat | hits | average |
| 2 | Thomas | 399 | 141 | |
| 3 | Franco | 433 | 138 | |
| 4 | Jackson | 369 | 115 | |
| 5 | Guillen | 365 | 105 | |
| 6 | Ventura | 401 | 113 | |
| 7 | LaValliere | 139 | 39 | |
| 8 | L. Johnson | 412 | 114 | |
| 9 | Cora | 312 | 86 | |
| 10 | Martin | 131 | 36 | |
| 11 | Raines | 384 | 102 | |
| 12 | Karkovice | 207 | 44 | |

## SPIRAL REVIEW    Chapter 1

**For Exercises 1–4, use the spreadsheet below.**

| | A | B | C | D | E | F | G |
|---|---|---|---|---|---|---|---|
| | | | CARSALES.SS | | | | |
| 1 | | Consumer | Business | Government | % Consum | % Bus | %Gov |
| 2 | 1993 | 4672 | 3943 | 100 | | | |
| 3 | 1992 | 4558 | 3683 | 113 | | | |
| 4 | 1991 | 4538 | 3752 | 97 | | | |
| 5 | 1990 | 5768 | 3567 | 149 | | | |
| 6 | 1989 | 6375 | 3402 | 136 | | | |
| 7 | | | | | | | |

**1.** Write a matrix of the data in the spreadsheet.

**2.** What are the dimensions of the matrix?

**3.** Write a variable expression and then a spreadsheet formula for finding the percentage of the automobiles sold in 1992 that were sold to businesses.

**4.** How do the expression and the spreadsheet formula in Exercise 3 compare?

**5.** The ages of the U.S. presidents at the time they took the oath of office (the first time if serving more than one consecutive term) are listed below. Find the mean, the median, and the mode of the data.

57, 61, 57, 57, 58, 57, 61, 54, 68, 51, 49, 64, 50, 48, 65, 52, 56, 46, 54, 49, 50, 47, 55, 55, 54, 42, 51, 56, 55, 51, 54, 51, 60, 62, 43, 55, 56, 61, 52, 69, 64, 46

**6. Writing** If you were going to present the data in Exercise 5 in a graph, would you use a bar graph or a histogram? Explain.

**Evaluate each variable expression for $x = -2$.**

**7.** $3x^2 + 2x - 4 + x^2$

**8.** $x + 3x + 5 + 2x^2 + 7$

# Section 2.1 Solving One-Step Equations

**GOAL**

**Learn how to . . .**
- write and solve one-step equations

**So you can . . .**
- solve real-world problems

### Application

In 1993, 3.5 million Japanese tourists spent $13.7 billion in the United States. An equation can be used to find the average amount spent by each Japanese tourist.

$$3{,}500{,}000x = 13{,}700{,}000{,}000$$
$$\uparrow$$
$$\text{average}$$

### Terms to Know | Example / Illustration

| Terms to Know | Example / Illustration |
|---|---|
| **Equation (p. 55)** — a mathematical sentence that says two numbers or expressions are equal | $x + 2 = 10$ |
| **Solve an equation (p. 55)** — finding the value of a variable that makes the equation true | The equation $x + 2 = 10$ can be solved by subtracting 2 from both sides of the equation. $$x + 2 = 10$$ $$x + 2 - 2 = 10 - 2$$ $$x = 8$$ |
| **Solution (p. 56)** — any value of a variable that makes an equation true | The solution of the equation $x - 7 = 6$ is $x = 13$. |
| **Inverse operations (p. 56)** — operations that undo each other | The inverse operation for division by 3 is multiplication by 3. |

## UNDERSTANDING THE MAIN IDEAS

### Inverse operations

The goal in solving equations is to get the variable alone on one side of the equation. Inverse operations are used to accomplish this goal. The operations of addition and subtraction undo each other. For example, adding 5 is undone by subtracting 5. The operations of multiplication and division undo each other.

### Example 1

Solve each equation.

**a.** $x + 4 = 10$

**b.** $-10 = x - 5$

**c.** $10 = \dfrac{x}{2}$

**d.** $3x = 7.5$

### Solution

**a.** To get $x$ alone, undo the addition.

$x + 4 = 10$
$x + 4 - 4 = 10 - 4$ ← *Subtract* 4 from both sides of the equation.
$x + 0 = 6$ ← $4 - 4 = 0$
$x = 6$

**b.** To get $x$ alone, undo the subtraction.

$-10 = x - 5$
$-10 + 5 = x - 5 + 5$ ← *Add* 5 to both sides of the equation.
$-5 = x + 0$ ← $-5 + 5 = 0$
$-5 = x$

**c.** To get $x$ alone, undo the division.

$10 = \dfrac{x}{2}$

$(10)(2) = \left(\dfrac{x}{2}\right)(2)$ ← *Multiply* both sides of the equation by 2.

$(10)(2) = (x)\left(\dfrac{2}{2}\right)$

$20 = x \cdot 1$ ← $\left(\dfrac{2}{2}\right) = 1$

$20 = x$

**d.** To get $x$ alone, undo the multiplication.

$3x = 7.5$

$\dfrac{3x}{3} = \dfrac{7.5}{3}$ ← *Divide* both sides of the equation by 3.

$(x)\left(\dfrac{3}{3}\right) = 2.5$

$x \cdot 1 = 2.5$ ← $\dfrac{3}{3} = 1$

$x = 2.5$

---

**Solve each equation.**

**1.** $x + 6 = 10$

**2.** $4 = x - 5$

**3.** $-11 + x = 7$

**4.** $4m = 120$

**5.** $\dfrac{r}{4} = 16$

**6.** $-18 = \dfrac{r}{5}$

## Opposites

Thinking about opposites will help you solve some equations. The opposite of $x$ is $-x$. Read $-x = 3$ as, "The opposite of $x$ is 3." Read $x = -3$ as, "$x$ is the opposite of 3."

> **Example 2**
>
> Solve the equation $-7 = -x + 3$.
>
> ■ **Solution** ■
>
> First, subtract 3 from both sides of the equation.
>
> $-7 = -x + 3$
> $-7 - 3 = -x + 3 - 3$
> $-10 = -x$  ← $-7 - 3 = -7 + (-3) = -10$
>
> So, $-10$ is the opposite of $x$ (or $x$ is the opposite of $-10$).
>
> Therefore, the value of $x$ is 10; that is, the solution of the equation is $x = 10$.

**Solve each equation.**

7. $-c + \dfrac{1}{8} = \dfrac{1}{2}$
8. $-\dfrac{x}{3} = -0.25$
9. $-8u = 48$
10. $-v = 0$
11. $5 = 8 - s$
12. $-n + 6 = 15$

................
### Spiral Review

**Evaluate each expression for $n = -2$.** *(Sections 1.4 and 1.5)*

13. $4n + 6$
14. $5 - 6n$
15. $3n + \dfrac{8}{2}$
16. $|5n|$
17. $\dfrac{|x + 3|}{6}$
18. $|-n| + |n|$

**GEOMETRY** Write a variable expression for the perimeter of each figure. Evaluate each expression for $x = 5$. *(Section 1.6)*

19.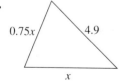

20.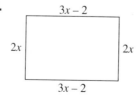

# Section 2.2 — Solving Two-Step Equations

**GOAL**

Learn how to . . .
- write and solve two-step equations

So you can . . .
- make purchasing decisions in real-world situations
- solve problems in geometry

## Application

As co-captains of the swim team, Daniel and Eric have the responsibility of choosing gifts for the teams' head coach and two assistant coaches. They have $75 to spend and have chosen a $35 sweatshirt for the head coach. They plan to spend an equal amount on each of the two assistant coaches. A two-step equation can be used to solve this problem.

$$2x + 35 = 75$$

## UNDERSTANDING THE MAIN IDEAS

The method for solving a two-step equation is the same as that for solving a one-step equation except that two inverse operations are required instead of one. The goal is still to get the variable alone on one side of the equation.

### Example 1

Solve each equation.

a. $3m - 2.5 = 1.5$

b. $\dfrac{w}{4} - 9 = 18$

**Solution**

a. **Step 1** Undo the subtraction by adding 2.5 to both sides.

$$\begin{aligned} 3m - 2.5 &= 1.5 \\ +2.5 &\phantom{=} +2.5 \\ \hline 3m &= 4 \end{aligned}$$

**Step 2** Undo the multiplication by dividing both sides by 3.

$$\dfrac{3m}{3} = \dfrac{4}{3}$$

$$m = \dfrac{4}{3}$$

**Check** Substitute $\dfrac{4}{3}$ for $m$.

$$3m - 2.5 = 1.5$$
$$3\left(\dfrac{4}{3}\right) - 2.5 \stackrel{?}{=} 1.5$$
$$4 - 2.5 \stackrel{?}{=} 1.5$$
$$1.5 = 1.5 \checkmark$$

*(Solution continued on next page.)*

Study Guide, ALGEBRA 1: EXPLORATIONS AND APPLICATIONS

## Solution (continued)

**b. Step 1** Undo the subtraction by adding 9 to both sides.

$$\frac{w}{4} - 9 = 18$$
$$\underline{\quad +9 \quad +9\quad}$$
$$\frac{w}{4} = 27$$

**Step 2** Undo the division by multiplying both sides by 4.

$$4\left(\frac{w}{4}\right) = 4(27)$$
$$w = 108$$

**Check** Substitute 108 for $w$.

$$\frac{w}{4} - 9 = 18$$
$$\frac{108}{4} - 9 \stackrel{?}{=} 18$$
$$27 - 9 \stackrel{?}{=} 18$$
$$18 = 18 \checkmark$$

**Solve each equation.**

1. $10 + 2d = 15$
2. $6 + 4r = -12$
3. $28 = 8 - 4h$
4. $24 = 3z - 9$
5. $-8 - 2x = 17$
6. $-6 - 5n = -15$
7. $\frac{x}{7} - 2 = -14$
8. $9 = \frac{x}{4} + 6$
9. $\frac{w}{8} - 3 = 11$
10. $5 = \frac{c}{3} - 6$
11. $-9 + \frac{m}{5} = 2$
12. $-3 = 5 + \frac{u}{6}$

## Example 2

For her birthday, four of Michi's friends have decided to buy a joint present and to take her to a movie. Each friend will pay an equal share of both the present and Michi's movie ticket, plus $6.50 for their own ticket. They decide to spend $15 each. How much can they spend in all on Michi's present and her ticket?

## Solution

*Problem Solving Strategy:* Use an equation.

Let $a$ = the amount they can spend on Michi's present and movie ticket. *Note:* This is the amount to be shared equally.

$$\frac{a}{4} + 6.5 = 15$$

Solve this equation for $a$.   *(Solution continued on next page.)*

> **Solution** (continued)
>
> **Step 1** Subtract 6.5 from both sides of the equation.
>
> $$\frac{a}{4} + 6.5 = 15$$
> $$\phantom{\frac{a}{4}} -6.5 \quad -6.5$$
> $$\frac{a}{4} = 8.5$$
>
> **Step 2** Multiply both sides of the equation by 4.
>
> $$4\left(\frac{a}{4}\right) = 4(8.5)$$
> $$a = 34$$
>
> So, the four friends can spend a total of $34 on Michi's present and movie ticket.

13. A tortoise and a hare both left the same spot at the same time. The hare traveled at a speed of 30 mi/h. After 30 min, the hare was 14.915 mi ahead of the tortoise. How fast was the tortoise traveling?

14. Rama has $250 with which to buy fencing for a dog pen. The fencing is sold in 8-foot sections at a price of $17.80 each. Rama also needs two 3-foot gates which sell for $32 each. How many 8-foot fence sections can she buy?

15. **Open-ended Problem** Use your answer from Exercise 14 to find the perimeter of Rama's dog pen. On centimeter grid paper, draw and label the pen. *Note*: Remember that the fence sections cannot be divided.

## Spiral Review

**Write a variable expression to describe each situation. Tell what the variable stands for.** *(Section 1.6)*

16. Today's high temperature is 15° colder than yesterday's high temperature.

17. Jose earns $8.75 an hour.

18. The ninth grade enrollment is up 17 students from last year.

**Simplify each expression.** *(Section 1.7)*

19. $2x + 5x - 10$
20. $6(-8 - t)$
21. $-3(5n - 6)$
22. $4 - 2(5 + 6w)$
23. $10 - (3r + 7)$
24. $5(4x + 3) - 3(x + 5)$

# Section 2.3 Applying Functions

**GOAL**

**Learn how to . . .**
- recognize and describe functions using tables and equations

**So you can . . .**
- predict the outcome of a decision

## Application

A pet store sells gerbils for $6.99 each and gerbil food for $1.09 per pound. The total amount you spend is a function of the number of pounds of food you buy.

| Input<br>Pounds of gerbil food | Output<br>Total cost (dollars) |
|---|---|
| 0 | 6.99 + 1.09(0) = $6.99 |
| 1 | 6.99 + 1.09(1) = $8.08 |
| 2 | 6.99 + 1.09(2) = $9.17 |
| 3 | 6.99 + 1.09(3) = $10.26 |

### Terms to Know / Example / Illustration

| Terms to Know | Example / Illustration |
|---|---|
| **Function (p. 66)**<br>relationship between input and output | In the equation $P = 5m$, the value of $P$ (the output) is a function of the value of $m$ (the input). |
| **Domain (p. 67)**<br>all input values that make sense for a function | Depending on the situation, the domain of the function $P = 5m$ might be the set of whole numbers. |
| **Range (p. 67)**<br>all possible output values | If the domain of the function $P = 5m$ is all the whole numbers, then the range is 0, 5, 10, 15, ... . |

## UNDERSTANDING THE MAIN IDEAS

### Functions

A function has exactly one output for each input. Two ways of representing a function are in a table or by an equation.

## Example 1

If the student council sells at least 100 magazine subscriptions, they make $0.50 profit on each subscription plus a bonus of $25. Their profit is a function of the number of magazines sold. Letting $P$ = the profit and $n$ = the number of magazines sold, the function is $P = 0.5n + 25$.

**a.** Make a table of input values and their corresponding output values for the profit function $P = 0.5n + 25$.

**b.** Suppose the student council made a profit of $232 from the subscription sales. How many subscriptions did they sell?

### ■ Solution ■

**a.** The input values (the number of subscriptions sold) are the whole numbers greater than or equal to 100.

| Input<br>Subscriptions sold | Output<br>Profit (dollars) |
|---|---|
| 100 | $0.5(100) + 25 = \$75.00$ |
| 101 | $0.5(101) + 25 = \$75.50$ |
| 102 | $0.5(102) + 25 = \$76.00$ |
| 103 | $0.5(103) + 25 = \$76.50$ |
| ... | ... |

← These three dots show that the pattern of numbers continues.

**b.** Use the equation $P = 0.5n + 25$. To find the number of subscriptions, substitute 232 for $P$ and find the value of $n$.

$$232 = 0.5n + 25$$
$$\underline{-25 \qquad -25} \quad \leftarrow \text{Subtract 25 from both sides.}$$
$$207 = 0.5n$$
$$\frac{207}{0.5} = \frac{0.5n}{0.5} \quad \leftarrow \text{Divide both sides by 0.5.}$$
$$414 = n$$

The student council sold 414 magazine subscriptions.

**Write an equation for each situation. Then make a table of values for the function.**

1. The senior class is selling yearbooks for $18.50 each. The revenue generated by the sales is a function of the number of yearbooks sold.

2. Ms. Snowe is an educational consultant to school districts. She charges a travel allowance of $0.24 per mile when she uses her car to drive to a school district plus a fee of $100.00 per day. Her total charge for a one-day trip to a school district is a function of the number of miles she drove to get there and back home.

**Use your equation from Exercise 1.**

3. Victor has collected $314.50 from the sale of yearbooks. How many yearbooks has he sold?

4. Erika has sold 47 yearbooks. How much money should she have collected?

5. Anoki has sold 3 more yearbooks than Erika. How much money should he have collected?

**Use your equation from Exercise 2.**

6. Last Tuesday, Ms. Snowe traveled a total of 257 mi to give a talk at Stanhope High School. What were her charges for that one-day trip?

7. Ms. Snowe charged the Frisco Community Schools $135.76 for a consulting session last Friday. How many miles did she travel that day?

8. A round-trip airline ticket to a city 450 mi from Ms. Snowe's residence costs $298. Which is more reasonable for that city's school system, to fly Ms. Snowe to their city to give her presentation or to pay her mileage for the use of her car to drive there? Explain.

## *Domain and range*

The domain of a function consists of any reasonable input values. The range is all the output values resulting from the input values; that is, the range depends on the domain.

### Example 2

Describe the domain and range of the function in Example 1.

### ▪ Solution ▪

The domain consists of the whole numbers 100, 101, 102, 103, ... . *Note*: The whole numbers less than 100 are not part of the domain because the equation is for profits on sales of *at least* 100 subscriptions. The range consists of these amounts in dollars and cents: 75.00, 75.50, 76.00, 76.50, ... .

**Match each equation with a situation. Tell what the variables stand for.**

**A.** $C = 150n$  **B.** $C = g + 0.50$  **C.** $C = \dfrac{l}{3} + 1.70$  **D.** $C = 0.06p$

9. The sales tax rate is 6%. The amount of sales tax is a function of the price of the item purchased.

10. In 1992, stores charged an average of 50 cents more per pound for apples than the amount they paid the growers. The price charged by stores is a function of the price paid to the growers plus the markup.

11. The rental charge for a vacation condominium is $150 night. The price paid is a function of the number of nights for which the condominium is rented.

12. Three friends decide to divide their lunch bill equally. They each paid $1.70 to ride the subway to the restaurant. The cost for each person is a function of their share of the lunch plus the cost of the subway ride.

## Spiral Review

**Solve each equation.** *(Sections 2.1 and 2.2)*

13. $22 = -x$
14. $-7 + x = -12$
15. $36 - x = -12$
16. $3x = 52.5$
17. $18 = 6 + 3x$
18. $-33 = -4x - 5$
19. $\dfrac{x}{14} = 9$
20. $\dfrac{x}{5} - 12 = -8$
21. $\dfrac{x}{4} + 7 = 8$

# Section 2.4 Coordinate Graphs

**GOAL**

**Learn how to . . .**
- read and create coordinate graphs

**So you can . . .**
- analyze data

### Application

Do you know how stars are created? Photos from the Hubble Space Telescope show jets of gas propelled out of forming stars at speeds up to 300 miles per second. The graph at the right shows how far these jets can travel in a matter of seconds.

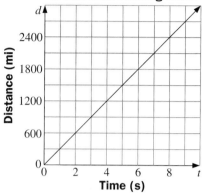

Distances Traveled by Jets of Gas from Forming Stars

### Terms to Know

### Example / Illustration

| Terms to Know | Example / Illustration |
|---|---|
| **Horizontal axis (p. 72)**<br>the number line that goes from left to right in a graph | The horizontal axis of the graph in the Applicationin |
| **Vertical axis (p. 72)**<br>the number line that goes from bottom to top in a graph | The vertical axis of the graph in the Applicaton shows the distance in miles. |
| **Coordinate plane (p. 73)**<br>a grid formed by two perpendicular number lines called *axes* | In the coordinate plane below, the origin, the *x*-axis, and the *y*-axis are shown. 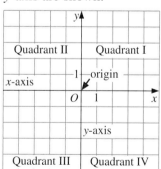 |
| **Origin (p. 73)**<br>the point with coordinates (0, 0) | |
| ***x*-axis (p. 73)**<br>the horizontal axis in a coordinate plane | |
| ***y*-axis (p. 73)**<br>the vertical axis in a coordinate plane | |

| Terms to Know | Example / Illustration |
|---|---|
| **Quadrant (p. 73)**<br>any of the four regions of a coordinate plane formed by the intersection of the *x*- and *y*-axes | In the coordinate plane at the bottom of the previous page, the quadrants are labled I, II, III, and IV. |
| **Ordered pair (p. 73)**<br>a pair of numbers used to identify a point in a coordinate plane | In the ordered pair (3, 900), 3 stands for the number of seconds and 900 stands for the distance in miles. |
| **Horizontal coordinate (p. 73)**<br>the first number in an ordered pair, also called the *x*-coordinate | In the ordered pair (2, 600), 2 is the horizontal coordinate. |
| ***x*-coordinate (p. 73)**<br>another name for a horizontal coordinate | In the ordered pair (6, 1800), the *x*-coordinate is 6. |
| **Vertical coordinate (p. 73)**<br>the second number in an ordered pair, also called the *y*-coordinate | In the ordered pair (5, 1500), the vertical coordinate is 1500. |
| ***y*-coordinate (p. 73)**<br>another name for a vertical coordinate | In the ordered pair (8, 2400), 2400 is the *y*-coordinate. |

## UNDERSTANDING THE MAIN IDEAS

### Applications

Coordinate graphs can be used to represent the relationship between two sets of data. The horizontal axis represents one set of data and the vertical axis represents the other set.

> **Example 1**
>
> Use the graph shown in the Application.
>
> **a.** How far can a jet of gas from a forming star travel in 4 s?
>
> **b.** The distance from New York to San Francisco is 2572 mi. About how long would it take a jet of gas from a forming star to travel this distance?

**Study Guide,** ALGEBRA 1: EXPLORATIONS AND APPLICATIONS
Copyright © McDougal Littell Inc. All rights reserved.

### Solution

**a.** Locate 4 on the horizontal axis. Move up this grid line until you reach the graphed line. Now move directly across to the left to find the corresponding value on the vertical axis.

A jet of gas from a forming star can travel 1200 mi in 4 s.

**b.** The distance 2572 mi is about halfway between 2400 mi and 2700 mi. So the location for 2572 on the vertical axis is about halfway between the grid lines labled "2400" and "2700." Move directly across to the right until you reach the graphed line. Now move directly down to find the corresponding value on the horizontal axis.

A jet of gas from a forming star can travel the distance from New York to San Francisco in about 8.5 s.

---

**For Exercises 1–4, use the graph shown in the Application.**

1. How long does it take a jet of gas from a forming star to travel 2700 mi?

2. How many miles does a jet of gas from a forming star travel in 6 s?

3. **Open-ended Problem** Think of the longest distance you have ever traveled and how long it took you to get there. Find out how long it would take the jet of gas from a forming star to travel that same distance.

4. The force behind the jets of gas is so powerful that astronomers believe the jets can travel trillions of miles into space. About how long would it take a jet of gas from a forming star to travel a million miles?

## *Graphing in the coordinate plane*

A coordinate plane is used to graph ordered pairs of numbers. It is formed by two perpendicular number lines, a horizontal line called the *x*-axis and a vertical line called the *y*-axis, that intersect at 0 on each number line. This intersection point is called the origin and it is represented by the ordered pair (0, 0). The *x*- and *y*-axes divide the coordinate plane into four regions called quadrants. The quadrants are numbered in a counterclockwise direction beginning with the uppper right quadrant (where the *x*- and *y*-coordinates of an ordered pair are both positive). Points that lie on either the *x*-axis or the *y*-axis are not in any quadrant.

### Example 2

Graph each point in a coordinate plane. Label each point with its letter. Name the quadrant (if any) in which the point lies.

    **a.** *B*(1, 3)             **b.** *C*(–2, –4)            **c.** *D*(0, –3)

## Solution

a. Start at the origin. The *x*-coordinate is positive so move 1 unit **right**. The *y*-coordinate is also positive, so now move 3 units **up**. Label the point *B*. The point is in Quadrant I.

b. Start at the origin. The *x*-coordinate is negative so move 2 units to the **left**. The *y*-coordinate is also negative, so now move 4 units **down**. Label the point *C*. The point is in Quadrant III.

c. Start at the origin. The *x*-coordinate is 0, so you do not move left or right. The *y*-coordinate is negative, so move 3 units **down**. Label the point *D*. The point is not in any quadrant.

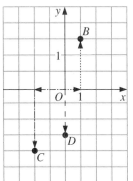

**Graph each point in a coordinate plane. Label each point with its letter. Name the quadrant (if any) in which the point lies.**

**5.** $A(1, 3)$    **6.** $B(-2, -1)$    **7.** $C(-4, 2)$    **8.** $D(0, -3)$

**9.** $E(2, -1)$    **10.** $F(4, 0)$    **11.** $G(-2.5, 0)$    **12.** $H\left(-\frac{7}{2}, -\frac{7}{2}\right)$

**For Exercises 13–16, use the figure at the right.**

**13.** Write an ordered pair for each point shown.

**14.** Name two points that lie in Quadrant II.

**15.** Name two points whose *x*-coordinate is greater than 0.

**16.** Name two points that do not lie in any quadrant.

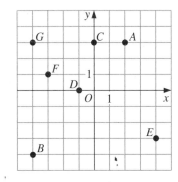

**PERSONAL FINANCE** Loans are given by banks to individuals for many uses, such as buying a car or house, or financing a college education. For Exercises 17–19, use the graph of monthly payments on a $5,000 loan shown at the right.

**17.** What does the horizontal axis show? What does the vertical axis show?

**18.** Estimate the difference between the monthly payments for a 1-year $5,000 loan and those for a 5-year $5,000 loan?

**19. Mathematics Journal** Zack wants to borrow $5,000. He has determined that he can afford a monthly payment of about $150. He is considering a 3- or 4-year loan length. Which do you think is the wiser choice? Explain.

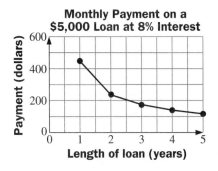

## Spiral Review

**For Exercises 20 and 21, find the mean, the median, and the mode of each set of data.** *(Section 1.3)*

20. Computer Games: $28.99, $29.99, $24.99, $29.99, $39.99, $39.99, $39.99, $43.99

21. World High Temperatures on July 10: 90, 59, 91, 86, 83, 91, 84, 93, 77, 66, 66, 91, 77, 84, 77

22. **Technology** Given the dimensions of each matrix, tell which matrices can be added. Then add them. Use a graphing calculator if you have one. *(Section 1.8)*

$$A = \begin{bmatrix} 1 & -3 & 7 \\ 6 & -4 & 8 \end{bmatrix} \qquad B = \begin{bmatrix} 2 & -5 & 4 \\ 7 & 9 & -3 \end{bmatrix} \qquad C = [-1 \ \ 6 \ \ -8]$$

$$D = \begin{bmatrix} 4 & 8 & 12 \\ 9 & -6 & -3 \\ 6 & 2 & -1 \end{bmatrix} \qquad E = \begin{bmatrix} 8 & -5 \\ -9 & 11 \\ 10 & 6 \end{bmatrix} \qquad F = [14 \ \ 27]$$

$$G = \begin{bmatrix} 7 & -9 \\ 8 & 1 \end{bmatrix} \qquad H = [12 \ \ 14 \ \ 16 \ \ 18] \qquad I = \begin{bmatrix} 5 & -7 \\ 6 & -8 \\ 4 & -3 \end{bmatrix}$$

# Section 2.5 Representing Functions

**GOAL**

**Learn how to . . .**
- graph equations

**So you can . . .**
- visually represent functions

## Application

The Music Boosters are considering selling sweatshirts from a pushcart at all extra-curricular activities. During the first year, they would make monthly payments of $43.50 on the pushcart while making a profit of $5.00 on every sweatshirt they sell. Using $P$ for profit and $s$ for sales, the equation $P = 5s - 43.5$ models the profit on sweatshirt sales. The president of the club made a graph like the one at the right above to show club members the possibilities for profit each month during the first year.

**Potential Sweatshirt Profits**

### Terms to Know

**Solution of an equation (p. 79)**
the ordered pair that makes an equation true

### Example / Illustration

The ordered pair (3, 1) is a solution of the equation $y = x - 2$.

## UNDERSTANDING THE MAIN IDEAS

You have learned that functions can be represented by equations and by tables of values. Graphs are another way to represent functions. Tables of values are used to draw graphs.

### Example 1

Make a table of values for the equation $y = x - 1$.

**Solution**

*Step 1* Set up a table with columns for the $x$- and $y$-coordinates. Write the equation above the table.

*Step 2* Choose at least three values of $x$. *Note*: Use both positive and negative numbers; also use values that can be graphed on a coordinate grid.

*Step 3* Find the corresponding $y$-value for each $x$-value.

| $y = x - 1$ | |
|---|---|
| $x$ | $y$ |
| −1 | −2 |
| 0 | −1 |
| 1 | 0 |

**Make a table of values for each function.**

1. $y = x + 8$
2. $y = 2x - 0.5$
3. $y = -\dfrac{x}{2}$

## Example 2

Use the table of values in Example 1 to graph the equation $y = x - 1$.

### Solution

*Step 1* Draw a coordinate grid. Be sure to make it large enough for the values in the table. Label the *x*- and *y*-axes and draw arrowheads to show their positive directions.

*Step 2* Plot the points from the table of values.

*Step 3* Using a straightedge, connect the points to make a line. Put an arrowhead on each end of the line to show that it extends in both directions. The completed graph is shown at the right.

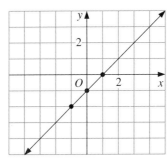

## Example 3

Using the graph in Example 2, identify two additional ordered pairs that are solutions of the equation.

### Solution

Any point on the line is a solution of the equation. For example, the points $(-2, -3)$ and $(2, 1)$ represent two solutions of the equation.

These solutions can be verified by substituting the values into the equation:

$-3 = -2 - 1$   and   $1 = 2 - 1$

**Graph the equation and the point. Tell whether the ordered pair is a solution of the equation.**

4. $y = -3x + 2$; $(2, 4)$
5. $y = -2 + x$; $(3, 1)$
6. $y = 3 - 2x$; $(2, -1)$
7. $y = \dfrac{x}{2}$; $(-4, 2)$
8. $y = x - 1$; $(3, 0)$
9. $y = -3 + x$; $(1, -2)$

**Match each equation with a graph.**

A. 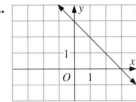   B. 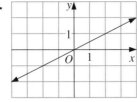   C.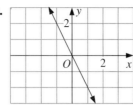

**10.** $y = -2x$     **11.** $y = 3 - x$     **12.** $y = 0.5x$

## Example 4

A new refrigerator costs $1000. Electricity to run the refrigerator costs about $68 per year. The total cost is a function of the number of years the refrigerator is used.

a. Write an equation for the function.

b. Make a table of values for the function.

c. Graph the function.

### Solution

a. Let $C$ = the total cost and $y$ = the number of years the refrigerator is used.

Since total cost = $68 per year + $1000, the equation is $C = 68y + 1000$.

b. Choose values for $y$, the number of years. *Note:* Think about the domain of the function, it must be the set of nonnegative numbers.

| Number of years | Total cost (dollars) |
|---|---|
| 0 | 1000 |
| 1 | 1068 |
| 2 | 1136 |
| 3 | 1204 |
| 4 | 1272 |
| ... | ... |

Use the equation $C = 68y + 1000$.

$$\begin{cases} 68(0) + 1000 = 1000 \\ 68(1) + 1000 = 1068 \\ 68(2) + 1000 = 1136 \\ 68(3) + 1000 = 1204 \\ 68(4) + 1000 = 1272 \end{cases}$$

c. Draw a pair of axes. Put the domain values on the horizontal axis and the range values on the vertical axis. Use the table from part (b) to plot at least three points. Since owners may keep refrigerators for part of a year, draw a line to connect the points. All points on the line are solutions of the equation. The completed graph is shown at the right.

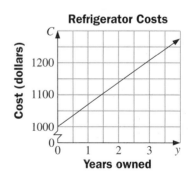

Refrigerator Costs

Study Guide, ALGEBRA 1: EXPLORATIONS AND APPLICATIONS
Copyright © McDougal Littell Inc. All rights reserved.

13. An air conditioner costs $280. Electricity to run the air conditioner costs 11.4¢ per kilowatt hour. The total cost of the air conditioner is a function of the number of kilowatt hours it is used.

   a. Write an equation for the function.

   b. Make a table of values for the function.

   c. Graph the function.

14. The cost of installing cable television is $20. The monthly charge is $37.50. The total cost of cable television is a function of the number of months it is used.

   a. Write an equation for the function.

   b. Make a table of values for the function.

   c. Graph the function.

## Spiral Review

**Simplify each expression.** *(Sections 1.2, 1.4, and 1.5)*

15. $\dfrac{6}{9} \cdot \dfrac{3}{5} + \dfrac{7}{30}$

16. $\dfrac{-6(3-7)}{8}$

17. $\dfrac{5}{2} - 7 + 8(3)$

**For Exercises 18 and 19, simplify each variable expression.** *(Section 1.7)*

18. $13m + 4n - 6 - 6m + 3$

19. $15 - 4(6x + 9)$

20. Show that $-10a + 4(2.5a + 7b)$ is equivalent to $28b$. *(Section 1.7)*

# Section 2.6 Using Graphs to Solve Problems

**GOAL**

**Learn how to . . .**
- use a graph to solve a problem

**So you can . . .**
- analyze the costs and profits of running a business, for example

## Application

The Student Council is considering selling fruit juice at the end of the school day. They plan to spend $16.74 for supplies to get started. They know they can make 15.3¢ on each cup they sell. A graph can be used to show how many cups they will need to sell before they can begin to make a profit.

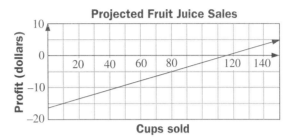

## UNDERSTANDING THE MAIN IDEAS

Graphs of equations can be used to solve problems and analyze data. A graphing calculator can help you make closer estimates than a hand drawn graph.

### Example

The Student Council decides that selling fruit juice could be profitable. They want to raise $2000 to attend a summer leadership conference. There are 185 days in a school year. About how many cups of juice will they need to sell each day in order to reach their goal?

### Solution

**Step 1** Write an equation. Let $P$ = profit over 185 days and $c$ = cups sold each day.

$$P = \underbrace{0.153c}_{\text{profit on one cup}} \cdot \underbrace{185}_{\text{days in school year}} - \underbrace{16.74}_{\text{initial costs}}$$

$P = 28.305c - 16.74$

*(Solution continues on next page.)*

**Study Guide,** ALGEBRA 1: EXPLORATIONS AND APPLICATIONS
Copyright © McDougal Littell Inc. All rights reserved.

### ■ Solution ■ (continued)

*Step 2* Make a table of values and graph the equation.

| $P = 28.305c - 16.74$ | |
|---|---|
| $c$ | $P$ |
| 10 | 266.31 |
| 20 | 549.36 |
| 30 | 832.41 |
| ... | ... |

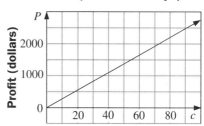

**Projected Fruit Juice Profits (185 Sales Days)**

*Step 3* Locate $2000 on the vertical axis and move across to the graphed line. Then move down to the horizontal axis to find the corresponding number of cups.

The Student Council will need to sell about 70 cups each day.

---

1. The Jones family is buying a washing machine that has yearly operating costs of $78 with an electric water heater and $31 with a gas water heater. They have an electric water heater. It would cost them $290 to buy a gas water heater. Write and graph an equation to show how long it would take the Jones to recover the cost of the gas water heater.

**The freshman class at a high school is paying a disc jockey $350 to play a dance. They are selling tickets to the dance for $3. Use this information for Exercises 2–4.**

2. Write and graph an equation that models the profit $P$ from selling $n$ tickets.

3. How many tickets must the class sell in order to break even? (To break even means they sell just enough tickets to pay the disc jockey.)

4. The class wants to make a profit of $200. How many tickets must they sell?

5. The 1995 movie *Apollo 13* cost $52 million to produce. Using an adult ticket price of $6.50, write and graph an equation to show how many adult tickets must be sold before the movie breaks even.

**TECHNOLOGY** For Exercises 6–8, graph each equation. Use a graphing calculator if you have one. Use the graph to estimate the missing value to the nearest tenth.

6. $y = 3.2x - 4$; find the value of $x$ when $y = 6.5$.

7. $y = 3 + 2.5x$; find the value of $x$ when $y = 7.4$.

8. $y = 5 - 3x$; find the value of $x$ when $y = -13.6$.

9. **Mathematics Journal** Describe how estimation is important in setting the viewing window on a graphing calculator.

## Spiral Review

Use the graph at the right. Tell whether each ordered pair is a solution of the equation $y = 4x + 2$. *(Section 2.5)*

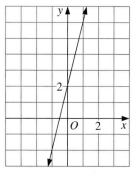

**10.** (2.5, 12)  **11.** (–1.5, –2)
**12.** (1, 6)  **13.** (–4, 14)

**Show three different ways to represent each function.** *(Section 2.5)*

**14.** Take any number, divide it by 4, and subtract 3.

**15.** Multiply any number by 3 and add 5 to it.

**16.** Take any number, add 3 to it, and divide by 2.

# Chapter 2 Review

**CHAPTER CHECK-UP**

Complete these exercises for a review of Chapter 2. If you have difficulty with a particular problem, review the indicated section.

1. Have you ever seen a $10,000 bill? In both 1993 and 1994 there were 3,450,000 in circulation. What is the total value of these bills? *(Section 2.1)*

**Solve each equation.** *(Sections 2.1 and 2.2)*

2. $4x = 75$
3. $4 = 5 + 2x$
4. $-x - 4 = -7$
5. $-x + \dfrac{1}{2} = \dfrac{2}{3}$
6. $-8 + \dfrac{x}{2} = 3$
7. $7 = \dfrac{x}{2}$

The charge for directory assistance calls is 40¢ for each call during a billing period. Use this information for Exercises 8–10. *(Section 2.3)*

8. Write an equation that describes the cost of directory assistance calls as a function of the number of calls made in a billing period.

9. Describe the domain and range of the function.

10. **Writing** Explain how this situation describes a function.

11. The graph at the right shows one family's monthly electrical costs over a one-year period. In what month did they spend the most on electricity? *(Section 2.4)*

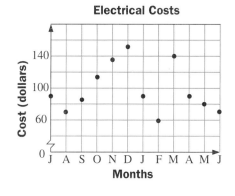

**Electrical Costs**

**Graph each point. Name the quadrant (if any) in which the point lies.** *(Section 2.4)*

12. $(-3, 2)$
13. $(4, 0)$
14. $(-2, -2)$
15. $(1, -3)$

**For Exercises 16–18, graph each equation in a coordinate plane.** *(Section 2.5)*

16. $y = x + 2$
17. $y = 3 - 2x$
18. $y = \dfrac{x}{0.5}$

19. Sloban spends $55 on supplies to make 125 hand-drawn greeting cards. She can make one card in about 20 min and spends about 3 h each day working on them. Sloban plans to sell the cards for $2.25 each. Write and graph an equation to find out how many cards she will need to sell in order to break even. *(Section 2.6)*

20. What will Sloban's profit be after she has sold all 125 cards? *(Section 2.6)*

21. **Open-ended Problem** Suppose Sloban wanted to make more money. Should she increase her production and sales or increase her price? Explain. *(Section 2.6)*

### SPIRAL REVIEW  Chapters 1–2

**Solve each equation.**

1. $-24 = 6x$
2. $-x - 5 = 3$
3. $\dfrac{x}{3} - 8 = 4$

**Use the graph at the right.**

4. Write two inequalities that describe the calorie content of the chips.

5. Find the mean, the median, and the mode of the data.

6. Choose any brand. Write an equation that describes the total number of calories in your chosen brand as a function of the number of ounces of chips. Describe the domain and range of the function.

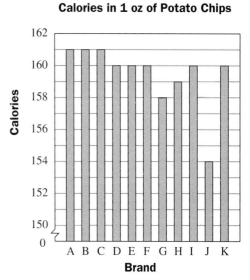

Calories in 1 oz of Potato Chips

**Graph each equation. Tell whether the ordered pair is a solution of the equation.**

7. $y = \dfrac{4x}{1.5} + 6;\ (-3, -2)$

8. $y = -4 + 2x;\ (-2, 0)$

**Evaluate each expression for $x = -2$.**

9. $\dfrac{|x - 3|}{x}$

10. $2(x - 5)$

11. $-3x - 8x$

**Write a variable expression for the perimeter of each figure. Evaluate each expression for $x = 7.5$.**

12.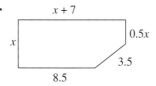

13.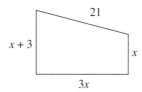

**Write a variable expression to describe each situation. Tell what the variable stands for.**

14. A pound of coffee costs $4.99.

15. Two-topping pizzas cost $3.75 more than one-topping pizzas.

**Simplify each expression.**

16. $-5 + 2x - 7 + 4 + x - 3$

17. $3(2x - 4) + 5(1.5x - 7)$

# Section 3.1 Applying Rates

**GOAL**

**Learn how to...**
- find unit rates from words and graphs

**So you can...**
- compare real-world rates

### Application

A small airplane flies 780 miles in 10 hours. What is the airplane's rate of speed for this flight?

| Terms to Know | Example / Illustration |
|---|---|
| **Rate (p. 99)** a ratio that compares two units | In the Application, the airplane's rate is $\frac{780 \text{ mi}}{10 \text{ h}}$. |
| **Unit rate (p. 99)** a rate per one given unit | Dividing 780 by 10 gives the airplane's unit rate in miles per hour: $\frac{78 \text{ mi}}{1 \text{ h}}$, or 78 mi/h. |

## UNDERSTANDING THE MAIN IDEAS

To convert a rate to a unit rate, divide the numerator by the denominator. You can convert rates to different units using conversion rates.

### Example 1

a. *Complete*: $\frac{\$10.50}{1.5 \text{ h}} = \frac{?}{1 \text{ h}}$

b. *Complete*: 200 m/h = _?_ km/h

c. *Complete*: 90 km/h = _?_ cm/s

### Solution

**a.** Dividing 10.50 by 1.5 gives 7.00; the rate is $\dfrac{\$7.00}{1\text{ h}}$.

**b.** Use the fact that 1 km = 1000 m.

$$\frac{200\text{ m}}{1\text{ h}} \times \frac{1\text{ km}}{1000\text{ m}} = \frac{200 \times 1\text{ km}}{1\text{ h} \times 1000} \quad \leftarrow \text{Cancel the labels ``m'' just like common factors.}$$

$$= \frac{1\text{ km}}{5\text{ h}}$$

$$= 0.2 \text{ km/h}$$

**c.** First use the fact that 1 km = 100,000 cm. Then use the fact that 3600 s = 1 h.

$$\frac{90\text{ km}}{1\text{ h}} \times \frac{100,000\text{ cm}}{1\text{ km}} = \frac{9,000,000\text{ cm}}{1\text{ h}}$$

$$= \frac{9,000,000\text{ cm}}{1\text{ h}} \times \frac{1\text{ h}}{3600\text{ s}}$$

$$= 2500 \text{ cm/s}$$

**Copy and complete each equation.**

1. $\dfrac{4\text{ people}}{1\text{ car}} \times \dfrac{10\text{ cars}}{1\text{ field trip}} = \underline{\ ?\ }$

2. $\dfrac{55\text{ pages}}{1\text{ chapter}} \times \dfrac{15\text{ chapters}}{1\text{ book}} = \underline{\ ?\ }$

3. $\dfrac{50\text{ mi}}{1\text{ h}} \times \underline{\ ?\ } = \dfrac{1200\text{ mi}}{1\text{ day}}$

4. $\dfrac{300\text{ ft}}{1\text{ s}} \times \underline{\ ?\ } = \dfrac{18000\text{ ft}}{1\text{ min}}$

**Express each rate in the given units. For conversion rates, see the Table of Measures on p. 612 of the Student Edition.**

5. 1200 mi/s = _?_ mi/min
6. $202 per week = _?_ per year
7. 1760 yd/mi = _?_ ft/mi
8. 65 mi/h = _?_ ft/min
9. 8.4 ft/s = _?_ mi/h
10. $5.50 per hour = _?_ cents per minute

### Example 2

Use the graph at the right to estimate the rate of change in the number of passenger cars in the world for the given years.

a. 1970–1975

b. 1980–1988

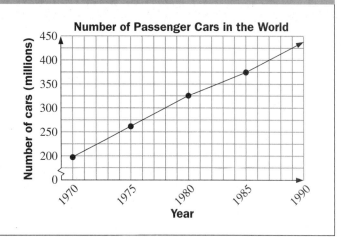

Number of Passenger Cars in the World

### ■ Solution ■

**a.** Use the estimates (1970, 200) and (1975, 260).

$$\text{rate of change} = \frac{\text{change in number of cars}}{\text{change in time}}$$

$$= \frac{260 - 200}{1975 - 1970}$$

$$= \frac{60}{5}$$

$$= 12$$

The rate of change during the years 1970–1975 was an increase of about 12 million cars per year.

**b.** Use the estimates (1980, 325) and (1988, 410).

$$\text{rate of change} = \frac{\text{change in number of cars}}{\text{change in time}}$$

$$= \frac{410 - 325}{1988 - 1980}$$

$$= \frac{85}{8}$$

$$= 10.625$$

The rate of change during the years 1980–1988 was an increase of about 10.625 million cars per year.

---

**For Exercises 11 and 12, use the graph at the right to estimate the rate of change in the number of pilots holding a commercial license for the given years.**

**11.** 1970–1980

**12.** 1980–1985

**13.** A newspaper article on the annual all-star football game reported that two quarterbacks combined for a record 312 passing yards and two touchdowns for the Gold team. Write two unit rates for this information.

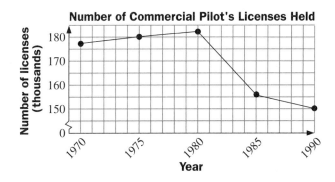

Number of Commercial Pilot's Licenses Held

**14.** For the population and area in square miles of the small states below, write a unit rate (in people per square mile) for each state.

| State | Population | Area (mi²) |
|-------|------------|------------|
| DE    | 666,168    | 8369       |
| CT    | 3,287,116  | 4873       |
| RI    | 1,003,464  | 1055       |
| VT    | 562,758    | 9271       |
| MA    | 6,016,425  | 7825       |

**15.** Explain what a comparison of unit rates tells you about the states listed in the table for Exercise 14.

**16. Mathematics Journal** Write a story based on the graph shown at the right. Be sure to include rates in your story.

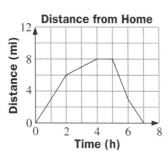

## Spiral Review

**Write each fraction as a decimal.** *(Toolbox, page 587)*

**17.** $\dfrac{6}{5}$      **18.** $\dfrac{3}{20}$      **19.** $\dfrac{13}{15}$      **20.** $\dfrac{7}{8}$

**Graph each point in a coordinate plane. Label each point with its letter. Name the quadrant (if any) in which the point is located.** *(Section 2.4)*

**21.** $R(-4, 3)$      **22.** $S(-2, -1)$      **23.** $T(3, -3)$

# Section 3.2

## Exploring Direct Variation

**GOAL**

Learn how to...
- recognize and describe direct variation

So you can...
- explore relationships between real-world variables

### Application

The table below shows the relationship between an astronaut's weight on Earth and the weight of the astronaut on the moon.

| Astronaut's Weights ||
|---|---|
| On Earth | On the Moon |
| 150 lb | 24.8 lb |
| 160 lb | 26.4 lb |
| 170 lb | 28.1 lb |
| 180 lb | 29.7 lb |

### Terms to Know

**Scatter plot (p. 107)**
a graph of points relating two sets of data

**Direct variation (p. 107)**
when two variables have a constant ratio

### Example / Illustration

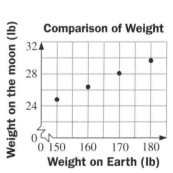

The ratio of

$$\frac{\text{astronaut's weight on the moon}}{\text{astronaut's weight on Earth}}:$$

$\frac{24.8}{150} \approx 0.165 \qquad \frac{26.4}{160} = 0.165$

$\frac{28.1}{170} \approx 0.165 \qquad \frac{29.7}{180} = 0.165$

Since the ratios are approximately the same, the relationship between weight on Earth and weight on the moon is a direct variation.

| **Constant of variation** (p. 107) the constant ratio of two variables in a direct variation relationship. | The constant of variation of the weights in the Application is approximately 0.165. |
|---|---|

## UNDERSTANDING THE MAIN IDEAS

There are two ways to test whether the relationship between two sets of data is a direct variation. You can calculate the ratio of all the data pairs, and if all the ratios are equal then a direct variation exists. You can also draw a scatter plot of the data, and if the relationship is a direct variation, there will be a line through the origin that passes through each point of the scatter plot.

A direct variation can be described by an equation in the form $y = kx$ or $\frac{y}{x} = k$, where $k \neq 0$ ($k$ represents the constant of variation). You can say that $y$ varies directly with $x$.

### Example

a. Tell whether this table shows direct variation.

| $x$ = distance in miles | 5 | 10 | 15 |
|---|---|---|---|
| $y$ = distance in kilometers | 8 | 16 | 24 |

b. For the data in the table above, give the constant of variation.

c. Write an equation for the data in the table.

### Solution

a. Compute all the ratios, $\frac{y}{x}$.

$$\frac{8}{5} = 1.6 \qquad \frac{16}{10} = 1.6 \qquad \frac{24}{15} = 1.6$$

Since all three ratios are the same, this is a direct variation.

b. From part (a), the constant of variation is 1.6.

c. Let $x$ = the distance in miles and $y$ = the distance in kilometers. Using the constant of variation found in part (b), the equation is:

$$\frac{y}{x} = 1.6, \text{ or } y = 1.6x$$

Tell whether the data show direct variation. If they do, give the constant of variation and write an equation that describes the situation.

**1.**

| Time (s) | Distance (ft) |
|---|---|
| 2 | 5.0 |
| 5 | 12.5 |
| 12 | 30.0 |

**2.**

| Age (yr) | Weight (lb) |
|---|---|
| 10 | 80 |
| 15 | 120 |
| 25 | 145 |

Decide whether each scatter plot suggests that the data show direct variation. Give reasons for your answers.

**3.**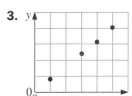

**4.**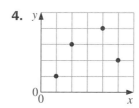

Leonardo da Vinci observed that the length of a person's face is $\frac{1}{9}$ of the person's height and a person's shoulder width is $\frac{1}{4}$ of the person's height. Use this information for Exercises 5–7.

**5.** Write an equation describing face length $f$ as a function of height $h$.

**6.** Write an equation describing shoulder width $w$ as a function of height $h$.

**7.** Does the length of a person's face vary directly with their shoulder width? Explain your answer.

**8.** The Supercalc Company determined that the cost of making solar calculators included $1500 per month for the building and machinery, plus $3.50 for each calculator made that month. Which of the graphs below shows the monthly cost for the company?

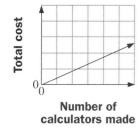
Number of calculators made

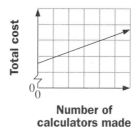
Number of calculators made

**9.** Does Supercalc's cost of making solar calculators vary directly with the number of calculators they make each month? Explain your answer.

**10.** Write an equation for the perimeter of each of the figures shown below. For each figure, decide whether the perimeter varies directly with *x*. Give reasons for your answers.

**Figure A**

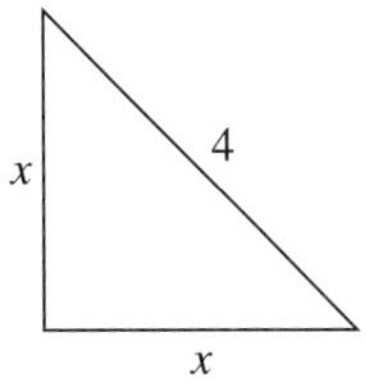

**Figure B**

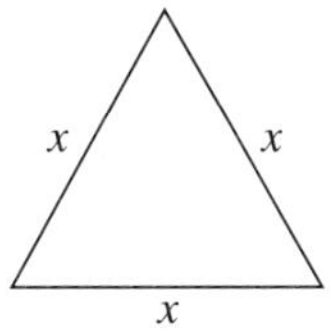

**11. Open-ended Problem** Draw two scatter plots, one that shows direct variation and another that does not.

## Spiral Review

**12.** Find the mean, the median, and the mode of data in the table at the right. *(Section 1.3)*

| Name | Hours worked per week |
|---|---|
| Reeta | 37.5 |
| Jon | 48.6 |
| Surya | 37.5 |
| Siobahn | 35.5 |

**Graph each equation and each point. Tell whether the ordered pair is a solution of the equation.** *(Section 2.5)*

**13.** $y = 3x - 1$; (3, 8)  **14.** $y = \frac{1}{3}x - 2$; (6, 4)

**15.** $y = 2 - 4x$; (−3, −10)  **16.** $y = 5x + 2$; (−2, −8)

## Section 3.3

# Finding Slope

**GOAL**

**Learn how to...**
- find the slope of a line

**So you can...**
- analyze a real-world graph

### Application

The graph at the right shows Jeff's distance (in miles) from home as a function of time (in hours) after leaving on a trip.

**Distance From Home**

*(Graph: Distance (mi) vs. Time (h); line rises from (0,0) through (4,160))*

### Terms to Know

### Example / Illustration

| Terms to Know | Example / Illustration |
|---|---|
| **Slope (p. 113)** the steepness of a line; a rate of change (The formula for slope is $\frac{\text{vertical change}}{\text{horizontal change}}$.) | For the graph in the Application, each horizontal change of 1 h results in a vertical change of 40 mi. The rate of change is 40 mi/h. |
| **Positive slope (p. 114)** the slope of a line that rises as you move from left to right | The line graphed in the Application has a positive slope. |
| **Negative slope (p. 114)** the slope of a line that falls as you move from left to right | The line graphed below has a negative slope. 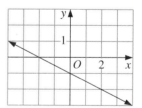 |
| **Undefined slope (p. 115)** the slope of a vertical line (The horizontal change of the line is 0.) | 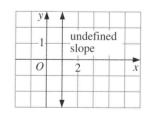 undefined slope |

Study Guide, ALGEBRA 1: EXPLORATIONS AND APPLICATIONS
Copyright © McDougal Littell Inc. All rights reserved.

# UNDERSTANDING THE MAIN IDEAS

To find the slope of a line from its graph, locate two points on the line and find the ratio of the vertical change to the horizontal change. Given two points $(x_1, y_1)$ and $(x_2, y_2)$ on a line,

$$\text{slope} = \frac{y_2 - y_1}{x_2 - x_1}$$

## Example 1

Find the slope of the line shown in the graph at the right.

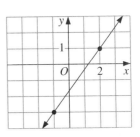

### Solution

The line passes through $(-1, -3)$ and $(2, 1)$. Let $(x_1, y_1) = (-1, -3)$ and $(x_2, y_2) = (2, 1)$.

$$\text{slope} = \frac{y_2 - y_1}{x_2 - x_1}$$

$$= \frac{1 - (-3)}{2 - (-1)}$$

$$= \frac{4}{3}$$

**Find the slope of each line.**

1.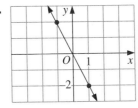
2.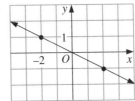
3.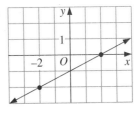

**Graph each equation. Find the slope of the line.**

4. $y = 4x - 1$
5. $y = 6 - x$
6. $y = -2$
7. $y = \frac{1}{3}x + 2$
8. $y = \frac{1}{2}x$
9. $y = -x - \frac{1}{4}$

### Example 2

Find the slope of the line that passes through the points (–3, 5) and (3, –5).

**Solution**

Let $(x_1, y_1) = (-3, 5)$ and $(x_2, y_2) = (3, -5)$.

$$\text{slope} = \frac{y_2 - y_1}{x_2 - x_1}$$

$$= \frac{-5 - 5}{3 - (-3)}$$

$$= \frac{-10}{6}, \text{ or } -\frac{5}{3}$$

**For exercises 10–15, find the slope of the line through each pair of points.**

**10.** (1, 4) and (–2, 2)  **11.** (0, 0) and (2, –4)  **12.** (2, 2) and (–5, 0)

**13.** (27, 15) and (18, 5)  **14.** (–8, 2) and (15, 2)  **15.** (3, –9) and (3, 17)

**16.** The American National Standards Institute specifications state that the maximum slope of a wheelchair ramp should be $\frac{1}{12}$. Does a ramp 5 in. high and 58 in. long meet this specification?

### Spiral Review

**Decide whether each table represents a function. Explain your reasoning.** *(Section 2.3)*

**17.**

| $x$ | Number whose square is equal to $x$ |
|---|---|
| 4 | 2 and –2 |
| 9 | 3 and –3 |
| 16 | 4 and 4 |

**18.**

| $x$ | $2.5x$ |
|---|---|
| 2 | 5 |
| 6 | 15 |
| 14 | 35 |

**Express each rate in the given units.** *(Section 3.1)*

**19.** 100 cm/s = _?_ km/h

**20.** $12 per hour = _?_ cents per minute

**21.** 30 ft/s = _?_ mi/h

**22.** 44 lb/min = _?_ tons/day

# Section 3.4 — Finding Equations of Lines

**GOAL**

**Learn how to...**
- write linear equations in slope-intercept form

**So you can...**
- analyze real-world problems shown by graphs

### Application

The graph at the right shows the cost $y$ of renting $x$ videos a year from SuperVid Rentals.

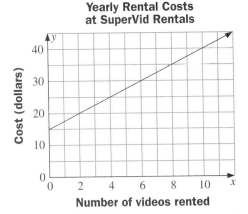

Yearly Rental Costs at SuperVid Rentals

### Terms to Know

| Terms to Know | Example / Illustration |
|---|---|
| **Linear equation (p. 119)**  an equation whose graph is a line | The equation $y = 2.5x + 15$ is a linear equation. Its graph is shown in the Application. |
| **Vertical intercept, y-intercept (p. 120)**  the y-value of the point where the graph crosses the y-axis | The y-intercept of the graph in the Application is 15. |
| **Slope-intercept form (p. 120)**  the form $y = mx + b$, where $m$ represents the slope of the line and $b$ represents the vertical intercept | The equation $y = 2x + 5$ is in slope-intercept form. |

### UNDERSTANDING THE MAIN IDEAS

You can write a linear equation in slope-intercept form from the graph of a line.
You can graph a line from the slope-intercept form of the equation.

**Study Guide,** ALGEBRA 1: EXPLORATIONS AND APPLICATIONS
Copyright © McDougal Littell Inc. All rights reserved.

## Example 1

Write an equation in slope-intercept form for the line graphed at the right.

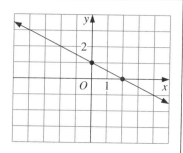

### Solution

Use the points $(0, 1)$ and $(2, 0)$ to find the slope.

$$m = \frac{0-1}{2-0} = -\frac{1}{2}$$

The line crosses the y-axis at $(0, 1)$, so the y-intercept is 1.

An equation of the line is $y = -\frac{1}{2}x + 1$.

**Write an equation in slope-intercept form for each graph.**

1.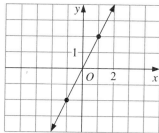
2.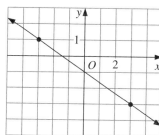
3.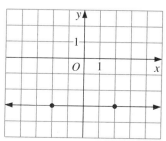
4.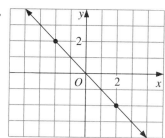
5.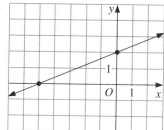
6.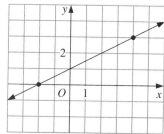

## Example 2

Graph the equation $y = \frac{3}{2}x - 2$.

### Solution

In slope-intercept form, the equation is $y = \frac{3}{2}x + (-2)$. From this equation, the slope of the line is $\frac{3}{2}$ and the y-intercept is $-2$.

Since the y-intercept is $-2$, begin by plotting the point $(0, -2)$. Now use the slope to find another point on the line. Since the slope is $\frac{3}{2}$, from the point $(0, -2)$ move 3 units up (the vertical change is +3) and 2 units right (the horizontal change is +2). Draw a dot at this point, $(2, 1)$. From this point, move 3 units up and 2 units to the right again. Draw another dot at this point, $(4, 4)$. Finally, draw a line through the three points. The completed graph is shown at the right.

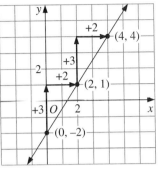

**For Exercises 7–12, graph each equation.**

7. $y = x - 2$
8. $y = 4x + 1$
9. $y = 2$
10. $x = -3$
11. $y = -\frac{1}{3}x - 2$
12. $y = \frac{5}{2}x + 3$

13. Corretta makes key rings to sell at craft fairs. The graph at the right shows her weekly costs. What does the vertical intercept of the line tell you? What does the slope of the line tell you?

14. **Mathematics Journal** Write a story that explains one family's costs for renting videos at SuperVid Rentals in one year. Use the information from the graph in the Application.

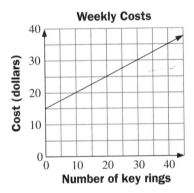

### Spiral Review

**Solve each equation.** *(Sections 2.1 and 2.2)*

15. $3x - 1 = 20$
16. $-3y - 8 = 16$
17. $-t + 3t = 26$
18. $\frac{1}{4}x + 7 = 11$
19. $\frac{2}{5}z - 1 = -5$
20. $-4w + 11w = 161$

# Section 3.5 Writing an Equation of a Line

**GOAL**

Learn how to...
- find the *y*-intercept of a graph

So you can...
- model real-world situations

*Application*

At the Grand Canyon, the August high temperatures vary with the elevation of the site. At the South Rim the elevation is 7050 ft and the high temperature is 92°F. At the North Rim the elevation is 8340 ft and the high temperature is 88°F. Assuming that the relationship is linear, the equation $t = -322.5e + 2,273,717$ expresses the high temperature as a function of the elevation.

## UNDERSTANDING THE MAIN IDEAS

To write the equation of a line in slope-intercept form, use the slope, *m*, and the *y*-intercept, *b*, to write the equation $y = mx + b$.

### Example 1

Find an equation of the line with slope 2 that passes through the point (3, –1).

**Solution**

Write the slope-intercept form with $m = 2$: $y = 2x + b$.

Now, substitute the coordinates of the point (3, –1) and solve for *b*.

$-1 = 2(3) + b$  ← Substitute 3 for *x* and –1 for *y*.

$-1 = 6 + b$

$-7 = b$

An equation of the line is $y = 2x - 7$.

**Find an equation of the line with the given slope and through the given point.**

1. slope = –1; (2, 5)
2. slope = 3; (4, 9)
3. slope = $\frac{1}{2}$; (6, 1)
4. slope = $-\frac{1}{3}$; (6, 3)
5. slope = 0; (–5, 2)
6. slope = –4; (1, –1)
7. undefined slope; (2, 3)
8. slope = 0.5; (5, –2)

### Example 2

Find an equation of the line through the points (0, –2) and (12, 10).

**Solution**

First, find the slope: $m = \dfrac{10 - (-2)}{12 - 0} = \dfrac{12}{12} = 1$.

Then substitute for $m$ in the equation: $y = 1x + b$.

Finally, substitute the coordinates of either point and solve for $b$.

$-2 = 1(0) + b$ ← Using (0, –2), substitute 0 for $x$ and –2 for $y$.
$-2 = b$

An equation of the line is $y = x - 2$.

**Find an equation of the line through the given points.**

9. (1, 6), (4, 3)
10. (1, 3), (4, 6)
11. (2, 8), (–4, 8)
12. (3, 1), (3, 12)
13. (0, –5), (3, 2)
14. (13, –2), (13, 3)
15. (–2, 6), (1, 0)
16. (0.3, –5), (6.3, –5)

### Spiral Review

**For Exercises 17–19, use the histogram at the right.** *(Section 1.1)*

17. How many students are at least 16 years old?
18. How many students are younger than 16 years old?
19. What percent of the students are over 15 years old?

20. A car rental company charges $20 per day and 25 cents per mile driven. Write a variable expression for the cost of driving this car $x$ miles in one day. *(Section 1.6)*

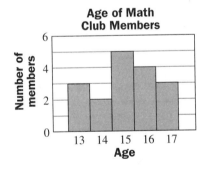

Age of Math Club Members

# Section 3.6

## Modeling Linear Data

**GOAL**

Learn how to...
- fit a line to data

So you can...
- make predictions about natural occurrences

### Application

The scatter plot below models the relationship between the number of marbles placed in a jar and the height of the water as given in the table.

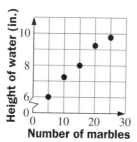

| Number of marbles in jar | Height of water in jar (in.) |
|---|---|
| 5 | 6 |
| 10 | 7.25 |
| 15 | 8.0 |
| 20 | 9.25 |
| 25 | 9.75 |

### Terms to Know

**Line of fit (p. 130)**
a line drawn on a scatter plot that is as close as possible to all of the points

### Example / Illustration

On the scatter plot below, a line of fit has been drawn.

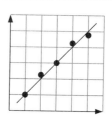

## UNDERSTANDING THE MAIN IDEAS

When the points on a scatter plot come close to forming a line, the data can be said to be linear. A line of fit can be used to model linear data.

### Example

a. Draw a line that looks as close as possible to all the points on the scatter plot shown at the right.

b. Choose any two points on your line of fit and write an equation of the line.

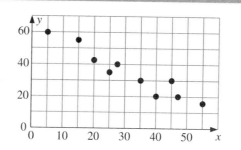

Study Guide, ALGEBRA 1: EXPLORATIONS AND APPLICATIONS
Copyright © McDougal Littell Inc. All rights reserved.

## ■ Solution ■

a. The line of fit shown in the figure at the right passes close to all the points. *Note*: There is not one correct placement of the line.

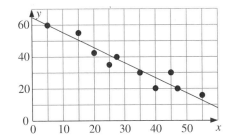

b. The line appears to pass through (5, 50) and (47, 20). The slope of the line is

$$m = \frac{20 - 50}{47 - 5} = \frac{-30}{42} = -\frac{5}{7}$$

So the equation of the line has the form $y = -\frac{5}{7}x + b$. Since (5, 50) is on the line, let $x = 5$ and $y = 50$.

$$50 = -\frac{5}{7}(5) + b$$

$$50 = -\frac{25}{7} + b$$

$$\frac{375}{7} = b$$

So the equation of the line is $y = -\frac{5}{7}x + \frac{375}{7}$.

**Find an equation that models the data in each scatter plot.**

1.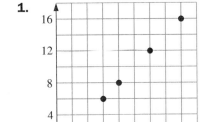

2.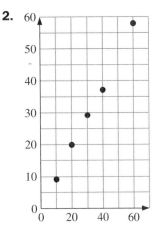

**The table shows the number of cricket chirps for various air temperatures. Use this data for Exercises 3–5.**

| Temperature (°F) | Cricket chirps per minute |
|---|---|
| 50 | 76 |
| 65 | 100 |
| 80 | 152 |
| 90 | 198 |

3. Make a scatter plot of the data. Draw a line of fit on your scatter plot. Find an equation of your line of fit.

4. Predict the number of cricket chirps per minute when the air temperature is 72°F.

5. The high temperature on a summer day reached 105°F. Use your equation from Exercise 3 to predict the number of cricket chirps per minute at this temperature.

The table shows a truck's maximum speed going uphill for various hill slopes. Use this data for Exercises 6–8.

| Slope of hill | Truck's maximum speed going uphill |
|---|---|
| 2° | 76 mi/h |
| 5° | 58 mi/h |
| 8° | 35 mi/h |
| 10° | 20 mi/h |

6. Make a scatter plot of the data. Draw a line of fit on your scatter plot. Find an equation of your line of fit.

7. Predict the truck's maximum speed going uphill when the slope of the hill is 6°.

8. The slope of a mountain highway is 12°. Use your equation from Exercise 6 to predict the truck's maximum speed going uphill.

## Spiral Review

**Simplify each variable expression.** *(Section 1.7)*

9. $16x - 2 - 12x + 5$
10. $2.5(4 - 10a)$
11. $12(2 + 1.5b)$
12. $2y - 3(y - 5)$
13. $-3(n - 6) + 7n$
14. $8(0.5x + 3)$

**Use a graph to find the solution.** *(Section 2.6)*

15. The choral music club has spent $450 on preparations for its spring concert. The club's goal is to make a profit of at least $600. If they sell tickets for $6 each, how many tickets does the club need to sell in order to reach its goal?

16. If the club spends $150 on advertising for its concert, how many tickets must be sold at $6 each to reach the club's goal of at least $600 profit?

# Chapter 3 Review

**CHAPTER CHECK-UP**

Complete these exercises for a review of Chapter 3. If you have difficulty with a particular problem, review the indicated section.

**For Exercises 1–4, copy and complete each equation.** *(Section 3.1)*

1. $\dfrac{150 \text{ mi}}{7 \text{ gal}} = \underline{\ ?\ }$ mi/gal

2. 176,825 mi/h ≈ $\underline{\ ?\ }$ mi/s

3. 48 ft/s = $\underline{\ ?\ }$ yd/min

4. 125 m/h = $\underline{\ ?\ }$ km/day

5. The weight of an astronaut on the moon varies directly with her weight on Earth. An astronaut who weighs 110 lb on Earth weighs 41.8 lb on the moon. What is the weight on Earth of an astronaut who weighs 47.5 lb on the moon? *(Section 3.2)*

6. **Writing** Explain how you can decide whether a graph represents a direct variation just by looking at it. *(Section 3.2)*

7. Find the slope of the line shown in the graph at the right. *(Section 3.3)*

8. Find the slope of the line through the points (–3, 3) and (5, –4). *(Section 3.3)*

9. Graph the equation $y = \dfrac{1}{2}x + 3$ and give the slope of the line. *(Section 3.3)*

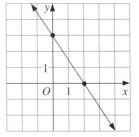

10. Write an equation in slope-intercept form for the graph shown at the right. *(Section 3.4)*

11. Graph the equation $y = \dfrac{5}{3}x - 1$. *(Section 3.4)*

12. Write an equation of the line with slope 0 that passes through the point (2, 5). *(Section 3.5)*

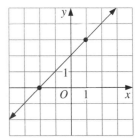

**Write an equation of the line through each pair of points.** *(Section 3.5)*

13. (0, 5) and (3, 12)

14. (5, 0) and (5, 12)

**For Exercises 15 and 16, refer to the table at the right.** *(Section 3.6)*

15. Make a scatter plot of the data in the table. Then draw a line of fit and write an equation of your line.

16. Use your equation to predict the suggested maximum weight for an adult 67 in. tall who is less than 35 years old.

| Suggested Maximum Weight for Adults Under 35 Years Old | |
|---|---|
| Height (in.) | Weight (lb) |
| 60 | 138 |
| 64 | 157 |
| 68 | 178 |
| 72 | 199 |

# SPIRAL REVIEW  Chapters 1–3

1. Convert $\frac{5}{11}$ to a decimal rounded to the nearest hundredth.
2. Tell whether $(-2, -8)$ is a solution of the equation $y = 2 - 5x$.
3. Tell whether the table below represents a function.

   | Height (in.) | 68 | 68 | 70 | 70 |
   |---|---|---|---|---|
   | Shoe size | 8 | 8.5 | 9 | 9.5 |

4. *Complete:* 5280 ft/min = _?_ mi/h
5. Solve $\frac{5}{6}x + 2 = -3$.
6. Write an equation for the graph at right that shows Sue's profit from the sale of paper flowers.
7. Use your equation from Exercise 6 to predict Sue's profit on a day when she sells 30 flowers.
8. Tell whether the table below represents a direct variation. Explain how you know.

   | Age of car (yr) | 1 | 2 | 3 |
   |---|---|---|---|
   | Total repair costs | $0 | $120 | $550 |

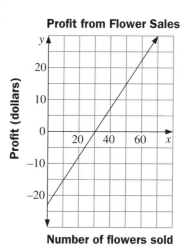

**Profit from Flower Sales**

Number of flowers sold

9. Gerie uses her car on her job. Her weekly salary is $600 and she gets paid $.29 per mile. Write an expression for her weekly pay, $P$, based on the number of miles, $m$, she drives.
10. The Computer Club spent $350 on games for the computer fair they will hold for the community. They will charge $3 admission, and hope to make a profit of $200. How many tickets will they need to sell to reach this goal?
11. Refer to Exercise 10. How much would the Computer Club need to charge for admission if they decide that they can only expect 100 people to attend?

# Section 4.1

## Solving Problems Using Tables and Graphs

**GOAL**

**Learn how to . . .**
- model situations with tables and graphs

**So you can . . .**
- compare options

### Application

The table below shows a comparison of the costs to rent videos at two video stores, one that charges a yearly membership fee and one that does not. Which video store would you rent videos from?

| Number of videos rented in a year | Cost at SuperVid | Cost at VidOne |
|---|---|---|
| 0 | $15 | $0 |
| 5 | $25 | $12.50 |
| 10 | $35 | $25 |
| 15 | $45 | $37.50 |
| 20 | $55 | $50 |

### UNDERSTANDING THE MAIN IDEAS

You can use tables and graphs to model situations involving comparisons, and then use the table or graph to choose the option that is best for you.

#### Example 1

a. Expand the table in the Application to 50 videos per year, and explain when you would choose to rent from SuperVid and when you would rent from VidOne.

b. For how many videos is the cost the same at the two stores?

---

**Study Guide,** ALGEBRA 1: EXPLORATIONS AND APPLICATIONS
Copyright © McDougal Littell Inc. All rights reserved.

### Solution

**a.**

| Number of videos | Cost at SuperVid | Cost at VidOne |
|---|---|---|
| 25 | $65 | $62.50 |
| 30 | $75 | $75.00 |
| 35 | $85 | $87.50 |
| 40 | $95 | $100.00 |
| 45 | $105 | $112.50 |
| 50 | $115 | $125.00 |

I would rent from SuperVid if I was renting more than 30 videos a year. I would rent from VidOne if I was renting less than 30 videos a year.

**b.** When 30 videos a year are rented, the cost is $75 from either store.

---

**1.** Customers with a regular checking account at BankOne pay a monthly fee of $5.50 per month plus $.10 per check. At People's Bank, there is a monthly fee of $4.00 for a regular checking account plus a charge of $.25 per check. The table below shows the total fee for writing checks each month at the two banks. Extend the table up to 10 checks.

| Number of checks | Fee at BankOne | Fee at People's Bank |
|---|---|---|
| 0 | $5.50 | $4.00 |
| 1 | $5.60 | $4.25 |
| 2 | $5.70 | $4.50 |

**2.** Which bank charges more if you write no checks during a month?

**3.** For how many checks is the total monthly fee the same at the two banks?

### Example 2

**Refer to the Application again.**

**a.** Write an equation for the cost $y$ of renting $x$ videos from SuperVid. Write an equation for the cost $y$ of renting $x$ videos from VidOne.

**b.** Graph the two equations from part (a). Explain how the graph can help you choose the best rental option for your family.

> ### ■ Solution ■
>
> **a.** At SuperVid, renting 5 videos costs $25, of which $15 is the membership fee. So the rental charge is $10 for 5 videos, or $2 per video. Therefore, the equation is $y = 15 + 2x$.
>
> At VidOne, the rental charge is $25 for 10 videos, or $2.50 per video. Therefore, the equation is $y = 2.50x$.
>
> **b.** The graph is shown at the right. It shows that the cost is lower at VidOne for $x$ less than 30, and higher for $x$ greater than 30. If a family rents fewer than 30 videos in a year, then they should rent from VidOne; if they plan to rent more than 30 videos in a year, they should rent from SuperVid.
>
>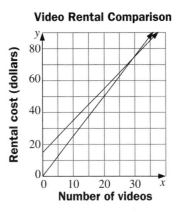

**For Exercises 4–7, refer to the situation described in Exercise 1.**

4. Write an equation that expresses the total fee $y$ for writing $x$ checks each month at BankOne.

5. Write an equation that expresses the total fee $y$ for writing $x$ checks each month at People's Bank.

6. Graph your equations from Exercises 4 and 5 on the same coordinate plane. What do the $y$-intercepts of the graphs tell you about each bank?

7. **Writing** Explain how you would decide which bank to use for your checking account.

**For Exercises 8–10, use the descriptions of two jobs given below.**

**Job A:** $245.00 each week plus $10 per hour for each hour of overtime.

**Job B:** $300.00 each week with no extra pay for overtime.

8. Use a graph to model the earnings $E$ from these two jobs as a function of the number of overtime hours $h$ worked each week.

9. How many hours of overtime would you have to work in a week to earn as much money from job A as from job B?

10. **Open-ended Problem** What questions would you ask in order to decide which job is better for you?

**Study Guide,** ALGEBRA 1: EXPLORATIONS AND APPLICATIONS
Copyright © McDougal Littell Inc. All rights reserved.

## Spiral Review

**Solve each equation.** *(Sections 2.1 and 2.2)*

**11.** $x - 12 = -7$        **12.** $\dfrac{m}{5} = 10$        **13.** $12 = \dfrac{b}{4} + 2$

**14.** $-3 = 2n + 13$        **15.** $17 = 4d - 11$        **16.** $23 - r = -31$

**Tell whether the data show direct variation. If they do, give the constant of variation and write an equation.** *(Section 3.2)*

**17.**

| Number of items sold | Profit |
|---|---|
| 10 | –$10 |
| 20 | $0 |
| 30 | $10 |

**18.**

| Operating time (min) | Number of parts produced |
|---|---|
| 30 | 226 |
| 120 | 902 |
| 270 | 2032 |

# Section 4.2 Using Reciprocals

**GOAL**

**Learn how to . . .**
- solve one-step and two-step equations using reciprocals

**So you can . . .**
- solve problems involving fractions

## Application
The area of a triangle is given by the formula $A = \frac{1}{2} \cdot b \cdot h$.

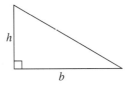

If $h = 3$ and $A = 5$, what is the value of $b$?

### Terms to Know | Example / Illustration

**Reciprocals** (p. 153)
two numbers whose product is 1

The numbers $\frac{2}{5}$ and $\frac{5}{2}$ are reciprocals because $\frac{2}{5} \cdot \frac{5}{2} = 1$ and $\frac{5}{2} \cdot \frac{2}{5} = 1$.

## UNDERSTANDING THE MAIN IDEAS

Equations that have a fraction as the coefficient of the variable can be solved by multiplying both sides of the equation by the reciprocal of the fraction.

### Property of Reciprocals

For every nonzero number $a$, there is exactly one number $\frac{1}{a}$ such that

$$a \cdot \frac{1}{a} = 1 \text{ and } \frac{1}{a} \cdot a = 1$$

### Example 1

Identify the reciprocal you would use to solve $\frac{2}{7}n = 12$.

■ **Solution** ■

Multiply by the reciprocal of $\frac{2}{7}$, the coefficient of $n$; the reciprocal of $\frac{2}{7}$ is $\frac{7}{2}$.

Identify the reciprocal you would use to solve each equation. Then solve.

**1.** $\frac{3}{5}x = 0$

**2.** $\frac{1}{2}n = 11$

**3.** $18 = \frac{5}{6}a$

---

**Example 2**

Rewrite the expression $\frac{-x}{7}$ with a fraction as the coefficient.

■ **Solution** ■

$\frac{-x}{7} = -\frac{1}{7}x$

---

Rewrite each expression with a fraction as the coefficient.

**4.** $\frac{5x}{12}$

**5.** $\frac{b}{10}$

**6.** $\frac{-2y}{19}$

---

**Example 3**

Solve the equation $\frac{-5x}{8} = 25$.

■ **Solution** ■

$\frac{-5x}{8} = 25$

$-\frac{5}{8}x = 25$

$\left(-\frac{8}{5}\right)\left(-\frac{5}{8}\right)x = \left(-\frac{8}{5}\right)(25)$

$x = -40$

---

Solve each equation.

**7.** $\frac{2}{3}x = 20$

**8.** $3 = -\frac{1}{6}n$

**9.** $-\frac{3}{5}y = -7$

**10.** $12 = \frac{7x}{2}$

**11.** $\frac{-13c}{2} = 26$

**12.** $16 = \frac{-4x}{3}$

## Spiral Review

**Simplify each expression.** *(Section 1.2)*

**13.** $9 - 3(16 - 5)$

**14.** $6 + 15 \div 3$

**15.** $\dfrac{21 - 6}{3} + 8(3)$

**Simplify each variable expression.** *(Section 1.7)*

**16.** $5x - 12 + x$

**17.** $0.23(5a - 3)$

**18.** $2(x + 1) + 11(x + 1)$

## Section 4.3 Solving Multi-Step Equations

**GOAL**

Learn how to...
- solve multi-step equations

So you can...
- solve real-world problems

### Application

The equation $y = 15 + 2x$ models the yearly cost for renting $x$ videos at SuperVid, while the equation $y = 2.50x$ models the cost at VidOne. Using these equations to determine the number of videos for which the cost is the same involves solving a multi-step equation.

### Terms to Know

**Identity** (p. 161)
an equation that is true for all numbers.

### Example / Illustration

The equation $5(x - 12) - 60 = 5x - 120$ is an identity because after using the distributive property and combining like terms on the left side, the resulting equation is $5x - 120 = 5x - 120$, which is true for any value of $x$.

### UNDERSTANDING THE MAIN IDEAS

To solve a multi-step equation with a variable term on both sides of the equals sign, follow these steps.

1. Simplify each side of the equation.
2. Get all the variable terms on one side of the equation. *Note:* Remember to keep the equation balanced by doing the same operation on both sides.
3. Get the variable alone on one side of the equation.

**Example 1**

Solve the equation $4x - 3 = x + 9$.

### Solution

Since both sides of the equation are simplified, the first step is to get the variable terms on one side of the equation. Then get the variable alone on that side of the equation.

$4x - 3 = x + 9$

$4x - 3 - x = x + 9 - x$ ← Subtract $x$ from both sides.

$3x - 3 = 9$ ← The variable term only appears on the left now.

$3x - 3 + 3 = 9 + 3$ ← Add 3 to both sides.

$3x = 12$ ← The variable term is alone on the left now.

$\frac{3x}{3} = \frac{12}{3}$ ← Divide both sides by 3.

$x = 4$

**Solve each equation.**

1. $6n = 2n + 16$
2. $5x - 2 = 3x + 8$
3. $4a - 10 = 3a - 1$
4. $5y - 6 = 8y + 6$
5. $7b - 5 = 5b - 7$
6. $9x + 5 = 4x$
7. $-3m + 4 = 5m + 4$
8. $6y - 17 = -3 - y$
9. $2r + 10 = 18 - 2r$

Some equations have *no solution*, because there is no value of the variable that will make the equation true. Sometimes *all numbers* are solutions of an equation. An equation that is true for all numbers is called an *identity*.

### Example 2

**Solve each equation.**

a. $2(3x - 5) = 3x + 5$
b. $2n - 5 = 3(n + 2) - n$

### Solution

a. First, simplify the left side.
Get both variable terms on one side.
Get $x$ alone on the left side.

$2(3x - 5) = 3x + 5$
$6x - 10 = 3x + 5$
$6x - 10 - 3x = 3x + 5 - 3x$
$3x - 10 = 5$
$3x - 10 + 10 = 5 + 10$
$3x = 15$
$\frac{3x}{3} = \frac{15}{3}$
$x = 5$

*(Solution continues on next page.)*

## Solution (continued)

b.
$$2n - 5 = 3(n + 2) - n$$
$$2n - 5 = 3n + 6 - n$$
$$2n - 5 = 2n + 6$$
$$2n - 5 - 2n = 2n - 2n + 6$$
$$-5 = 6$$

Since the statement $-5 = 6$ is never true, there are *no solutions* to this equation.

**Solve each equation. If an equation is an *identity* or there is *no solution*, say so.**

10. $4x - 5 = 4(x - 5)$
11. $-3(w + 5) = 12(1 - w)$
12. $13x = -(3x + 8)$
13. $4x - 20 = 4(x - 5)$
14. $2(3n - 6) = 3(2n - 4)$
15. $7(d - 5) = d + 1$

### Example 3

Refer to the Application. Use an equation to determine how many videos you would need to rent each year in order to make the total cost at SuperVid the same as the cost at VidOne.

## Solution

Set the two costs equal to each other and solve the equation.

$$15 + 2.00x = 2.50x$$
$$15 + 2.00x - 2.00x = 2.50x - 2.00x$$
$$15 = 0.50x$$
$$30 = x$$

You would have to rent 30 videos for the total costs to be the same.

16. One car rental company charges $20 per day plus $.25 per mile, while another company charges $25 per day plus $.20 per mile. For how many miles per day are the total costs at the two companies equal?

17. **Mathematics Journal** One job pays $220 per week plus $8 per hour of overtime over 40 hours per week, while another job pays $260 per week with no extra pay for overtime. Explain the method you would use to determine when you would choose the first job.

## Spiral Review

**For Exercises 18–20, find the least common multiple of each set of numbers.**
*(Toolbox, page 584)*

**18.** 3, 6, 15          **19.** 2, 4, 12          **20.** 3, 14, 18

**21.** Tobi collects returnable bottles and cans to bring to the recycling center. He receives $.05 for each bottle or can returned. The amount he receives is a function of the number of bottles and cans he returns. Find the domain and the range of the function. *(Section 2.3)*

# Section 4.4

## Equations with Fractions or Decimals

**GOAL**

**Learn how to . . .**
- solve equations that involve more than one fraction or decimal

**So you can . . .**
- solve real-world problems

### Application

A helicopter flies with the wind at a rate of 120 km/h from Peoria to Pekin and then returns to Peoria, flying against the wind, at a rate of 80 km/h. How far did the helicopter travel from Peoria to Pekin and back if the total trip took 30 min?

## UNDERSTANDING THE MAIN IDEAS

### Equations with fractions

To solve an equation with fractions, you can multiply every term in the equation by the least common denominator of the fractions. The least common denominator is the least common multiple (LCM) of the denominators of the fractions.

---

**Example 1**

What is the least common denominator of the fractions $\frac{7}{10}$, $\frac{2}{3}$, and $\frac{3}{5}$?

**Solution**

multiples of 3: 3, 6, 9, 12, 15, 18, 21, 24, 27, **30**, 33, 36, …

multiples of 5: 5, 10, 15, 20, 25, **30**, 35, 40, …

multiples of 10: 10, 20, **30**, …

The least common multiple of 10, 3, and 5 is 30. Therefore, the least common denominator of the fractions is 30.

---

**Find the least common denominator of each group of fractions.**

1. $\frac{1}{3}, \frac{1}{4}, \frac{4}{5}$

2. $\frac{2}{5}, \frac{3}{4}, \frac{1}{2}$

3. $\frac{2}{3}, \frac{5}{6}, \frac{12}{21}$

### Example 2

Solve the equation $\frac{1}{10}y + \frac{1}{3}y = \frac{3}{5}$.

### ■ Solution ■

The least common denominator of the fractions is 30, so begin by multiplying both sides of the equation by 30.

$$\frac{1}{10}y + \frac{1}{3}y = \frac{3}{5}$$

$$30\left(\frac{1}{10}y + \frac{1}{3}y\right) = 30\left(\frac{3}{5}\right) \quad \leftarrow \text{Use the distributive property.}$$

$$3y + 10y = 18$$

$$13y = 18 \quad \leftarrow \text{Combine like terms.}$$

$$\frac{13y}{13} = \frac{18}{13} \quad \leftarrow \text{Divide both sides by 13.}$$

$$y = \frac{18}{13}$$

**Solve each equation.**

4. $\frac{x}{3} + \frac{x}{5} = 4$

5. $\frac{m}{5} + \frac{m}{4} = \frac{18}{5}$

6. $\frac{3x}{5} - \frac{7}{10} = \frac{11}{10}$

7. $\frac{1}{3}n + \frac{3}{4}n = \frac{13}{3}$

8. $\frac{3}{5}p + \frac{1}{4}p = 17$

9. $\frac{3}{4}y + \frac{7}{8} = \frac{11}{8}$

10. $\frac{v}{3} + \frac{v}{7} = 20$

11. $\frac{b}{8} + \frac{b}{4} = 18$

12. $\frac{3a}{5} - \frac{1}{10} = \frac{11}{10}$

13. $\frac{1}{4}c + \frac{1}{5}c + \frac{5}{4} = \frac{9}{5}$

14. $\frac{3}{5}x - \frac{7}{10} = \frac{11}{10} + \frac{3}{10}x$

15. $\frac{5}{6}s - \frac{3}{4} = \frac{5}{8} + \frac{1}{3}s$

## *Equations with decimals*

To make an equation with decimals easier to solve, multiply both sides of the equation by a power of 10 that will eliminate all the decimals.

### Example 3

Solve the equation $16.2 - 4.5x = 3.2 - 0.5x$.

### Solution

Begin by multiplying both sides of the equation by 10 in order to eliminate the decimals.

$$16.2 - 4.5x = 3.2 - 0.5x$$
$$10(16.2 - 4.5x) = 10(3.2 - 0.5x)$$
$$162 - 45x = 32 - 5x \quad \leftarrow \text{Use the distributive property.}$$
$$162 - 45x + 5x = 32 - 5x + 5x \quad \leftarrow \text{Add } 5x \text{ to both sides.}$$
$$162 - 40x = 32 \quad \leftarrow \text{Combine like terms.}$$
$$162 - 40x - 162 = 32 - 162 \quad \leftarrow \text{Subtract 162 from both sides.}$$
$$-40x = -130$$
$$\frac{-40x}{-40} = \frac{-130}{-40} \quad \leftarrow \text{Divide both sides by } -40.$$
$$x = 3.25$$

**Solve each equation.**

**16.** $0.04x + 1.32 = 1.08$     **17.** $0.2m - 0.08 = 1.46$

**18.** $1.75n = 5 + 1.25n$     **19.** $5.5 - 0.01x = 2.24x - 3.5$

**20.** $4v - 0.65 = 2v - 0.4$     **21.** $5.4a - 0.004 = 0.05$

### Example 4

Refer to the Application. Find the distance that the helicopter traveled from Peoria to Pekin and back.

### Solution

Let $d$ = the distance from Peoria to Pekin.

Remember that $d = r \times t$ so $t = \frac{d}{r}$.

Traveling with the wind on the way to Pekin, $r = 120$ so $t = \frac{d}{120}$.

Returning against the wind, $r = 80$ so $t = \frac{d}{80}$.

Since the total round trip took $\frac{1}{2}$ h, the sum of the two times must be $\frac{1}{2}$.

*(Solution continues on next page.)*

> **■ Solution ■** *(continued)*
>
> $$\frac{d}{120} + \frac{d}{80} = \frac{1}{2}$$
>
> $$240\left(\frac{d}{120} + \frac{d}{80}\right) = 240\left(\frac{1}{2}\right) \quad \leftarrow \begin{array}{l}\text{The LCM of 2, 80, and 120 is 240;}\\ \text{multiply both sides by 240.}\end{array}$$
>
> $$2d + 3d = 120$$
>
> $$5d = 120 \quad \leftarrow \text{Combine like terms.}$$
>
> $$\frac{5d}{5} = \frac{120}{5}$$
>
> $$d = 24$$
>
> The distance from Peoria to Pekin is 24 km, so the round trip distance is 48 km.

**22.** Jon can paddle a canoe upstream at a rate of 8 km/h. When he makes the return trip with the current, he can paddle at a rate of 10 km/h. It takes Jon a half hour to make the trip upstream and back. Write an equation to find the total distance Jon paddled.

### Spiral Review

**TECHNOLOGY** For Exercises 23 and 24, graph each equation. Use a graphing calculator if you have one. Use the graph to estimate the unknown value to the nearest tenth. *(Section 2.6)*

**23.** $y = -x - 3$; find the value of $x$ when $y = 9$.

**24.** $y = \frac{1}{3}x - 2$; find the value of $x$ when $y = -2$.

**25.** Find the mean, the median, and the mode(s) of the data. *(Section 1.3)*

Days absent from school: 3, 4, 2, 2, 5, 4, 6, 5, 3, 3

# Section 4.5 Writing Inequalities from Graphs

**GOAL**

**Learn how to . . .**
- use inequalities to represent intervals on a graph

**So you can . . .**
- represent a group of solutions

### Application
A fitness club charges members $5 per visit after the member pays a $140 annual fee to join. Nonmembers pay $9 per visit to use the club. When is the total cost for a member of the club less than the total cost for a nonmember?

### Terms to Know | Example / Illustration

| Terms to Know | Example / Illustration |
|---|---|
| **Inequality (p. 171)** a mathematical statement formed by placing an inequality symbol between numerical or variable expressions | $5x + 140 < 9x$ <br> $14y \geq 38 - 5y$ |

## UNDERSTANDING THE MAIN IDEAS

When you solve an inequality graphically, the word "greater" indicates the part of the graph *above* some value; the word "less" indicates the part of the graph *below* some value.

### Example 1

a. Graph the line $y = x + 2$.

b. Write an inequality to represent the $x$-values when $y > 3$.

**Solution**

a. The equation $y = x + 2$ is in slope-intercept form. From the equation, the $y$-intercept of the line is 2 and the slope is 1. The graph of the equation is shown at the right.

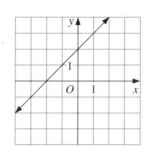

### ■ Solution ■

**b.** Find 3 along the *y*-axis and place a straightedge horizontally across the graph at this point. Now look for the part of the graphed line that is *above* (greater than) the straightedge. Note that this is the portion of the line for which *x* is greater than 1. Therefore, the inequality is $x > 1$.

**For Exercises 1–3:**

a. **Graph each equation.**
b. **Write an inequality to represent the *x*-values when *y* > 2.**

1. $y = 3x + 2$
2. $y = -3x + 2$
3. $y = 3x$

### Example 2

For the fitness club discussed in the Application, find the number of visits each year that will make a member's total cost less than a nonmember's total cost.

### ■ Solution ■

**Step 1** The total cost is a function of number of visits to the fitness club. Let $y$ = the total cost and $x$ = the number of visits. Then a member's total cost is modeled by the equation $y = 5x + 140$ and a nonmember's cost is modeled by the equation $y = 9x$.

**Step 2** Graph the two equations on the same coordinate plane.

**Step 3** Find the *x*-value of the point where the two graphs intersect: $x = 35$. So the *x*-values that make the graph of $y = 5x + 140$ lower than the graph of $y = 9x$ are the values $x > 35$.

Therefore, a member's total cost is less than that of a nonmember whenever they visit the club more than 35 times in one year.

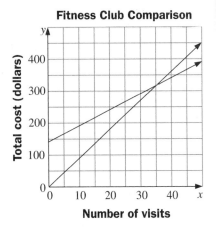

**Fitness Club Comparison**

The graph at the right shows the cost of buying CDs from two different music clubs. The cost is a function of the number of CDs a club member buys each year.

4. **Writing** Describe each club's cost in words.
5. When are the costs equal? Explain how you know.
6. When does it cost less to buy from ClubOne? Write your answer as an inequality.
7. When does it cost less to buy from Select-a-Club? Write your answer as an inequality.
8. **Writing** Explain how you would choose the club that you would buy from.

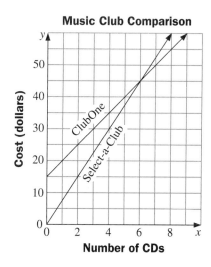

Music Club Comparison

## Spiral Review

**Find an equation of the line with the given slope and through the given point.** *(Section 3.5)*

9. slope = 0; (4, –3)
10. slope = $\frac{1}{2}$; (–12, 12)
11. undefined slope; (3, 3)
12. slope = 4; (–2, –5)

**Graph each inequality on a number line.** *(Toolbox, page 591)*

13. $x > 5$
14. $x \leq -2$
15. $n \leq 0$

# Section 4.6 Solving Inequalities

**GOAL**

**Learn how to . . .**
- solve inequalities

**So you can . . .**
- tell when one quantity is greater than another

## Application

It costs Elena $6.00 to see a movie at a theater near her home. She can rent a video for $2.50 per day. She would like to buy a VCR for $325 so that she can save money. How can she decide whether to buy the VCR?

### Terms to Know | Example / Illustration

**Solution of an inequality (p. 177)**
the set of numbers that makes an inequality true

$5x - 7 > 18$

$5x - 7 + 7 > 18 + 7$

$5x > 25$

$\dfrac{5x}{5} > \dfrac{25}{5}$

$x > 5$

The solution of the inequality $5x - 7 > 18$ is $x > 5$.

## UNDERSTANDING THE MAIN IDEAS

When you add, subtract, multiply, or divide both sides of an inequality by the same positive number, the inequality symbol stays the same. It also stays the same when you add or subtract the same negative number. When you multiply or divide both sides of an inequality by the same *negative* number, the inequality symbol is *reversed*.

### Example 1

Solve the inequality $5x + 1 < 3x - 3$. Graph the solution on a number line.

### Solution

$$5x + 1 < 3x - 3$$
$$5x + 1 - 1 < 3x - 3 - 1$$
$$5x < 3x - 4$$
$$5x - 3x < 3x - 4 - 3x$$
$$2x < -4$$
$$\frac{2x}{2} < \frac{-4}{2}$$ ← Dividing by a positive number does not reverse the inequality symbol.
$$x < -2$$

On a number line, graph all the numbers less than −2.

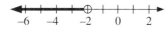

**Solve each inequality. Graph each solution on a number line.**

1. $x + 1 > 0$
2. $2x \geq 32$
3. $10 < m + 5$
4. $15y \leq 8$
5. $\frac{y}{-2} < 6$
6. $\frac{4}{5}n \geq 20$
7. $16 > -2x + 6$
8. $2 - y < 2$
9. $4x + 5 \leq -23$
10. $25 - 6c \geq 1$
11. $12 > 22 - 4b$
12. $2 + 3a > -10$

### Example 2

Solve the inequality $-3x \leq 12$. Graph the solution on a number line.

### Solution

$$-3x \leq 12$$
$$\frac{-3x}{-3} \geq \frac{12}{-3}$$ ← Dividing by a negative number reverses the inequality symbol.
$$x \geq -4$$

Graph −4 and all numbers greater than −4.

**Solve each inequality. Graph each solution on a number line. If there is no solution, write *no solution*.**

13. $-2a + 3 < 2a + 7$
14. $3y > 2(y - 9)$
15. $2(3y - 1) \leq 6y$

---

### Example 3

Refer to the Application. Find the answer to the question by solving an inequality.

**■ Solution ■**

Let $n$ = the number of movies that Elena watches. Then the cost of seeing the movies in the theater is $6.00n$ and the cost of buying a VCR and renting the videos is $325 + 2.50n$.

Elena needs to know when the cost of a VCR and renting videos will be less than paying to go to the movies.

$$325 + 2.50n < 6.00n \quad \leftarrow \text{cost of VCR} < \text{cost at theater}$$
$$325 + 2.50n - 2.50n < 6.00n - 2.50n$$
$$325 < 3.50n$$
$$92.9 < n$$

Elena must watch 93 movies before the cost of buying the VCR and renting videos is less than the cost of watching the movies at the theater.

---

16. Suppose that Elena decides to only go to the movies during the bargain matinee hours when the cost is $2.50. What is your advice to Elena now?

17. **Mathematics Journal** Write your own problem similar to the one in the Application. Show how to use an inequality to solve your problem.

..................
### Spiral Review

**Copy and complete each equation.** *(Section 3.1)*

18. $\dfrac{55 \text{ mi}}{1 \text{ h}} \cdot \dfrac{?}{\phantom{x}} = \dfrac{330 \text{ mi}}{1 \text{ day}}$

19. $\dfrac{\$21.00}{5 \text{ h}} \cdot \dfrac{1 \text{ h}}{60 \text{ min}} = \dfrac{?}{\phantom{x}}$

20. $\dfrac{2 \text{ dimes}}{1 \text{ toll}} \cdot \dfrac{\$1}{10 \text{ dimes}} \cdot \dfrac{8 \text{ tolls}}{1 \text{ day}} = \underline{\ ?\ }$

21. $\dfrac{32 \text{ ft}}{1 \text{ s}} \cdot \dfrac{?}{\phantom{x}} \cdot \dfrac{1 \text{ mi}}{5280 \text{ ft}} = \underline{\ ?\ } \text{ mi/h}$

**Find the slope of the line through each pair of points.** *(Section 3.3)*

22. $(2, -3)$ and $(2, 17)$
23. $(7, 1)$ and $(-2, -8)$
24. $(-5, 3)$ and $(4, 3)$

# Chapter 4 Review

**CHAPTER CHECK-UP**

Complete these exercises for a review of Chapter 4. If you have difficulty with a particular problem, review the indicated section.

For Exercises 1 and 2, use the two jobs decribed below.

Job A: $270 each week with no pay for overtime.
Job B: $230 each week plus $8.50 per hour for each hour of overtime.
*(Section 4.1)*

1. Use both a table and a graph to model the earnings $E$ from the two jobs as a function of the number of overtime hours $h$ worked each week.

2. How many hours of overtime would you have to work in a week in order to earn as much from job B as from job A?

For Exercises 3–13, solve each equation. *(Sections 4.2, 4.3, and 4.4)*

3. $\frac{4}{3}x = 28$

4. $5 = -\frac{1}{10}n$

5. $-\frac{2}{7}y = -5$

6. $-2x + 4 = 3x + 4$

7. $5y - 15 = -3 + y$

8. $3r + 2 = 3(8 + r)$

9. $\frac{2x}{3} + \frac{x}{4} = 22$

10. $\frac{3m}{5} + \frac{m}{4} = \frac{17}{5}$

11. $\frac{-5n}{6} + \frac{4n}{15} = 51$

12. $0.05x + 2.33 = 1.08$

13. $464 - 0.01m = -0.02m$

14. Graph the line $y = x - 2$. Write an inequality to represent the $x$-values when $y < 2$. *(Section 4.5)*

15. A fitness club charges as follows:
    members: $50 yearly membership fee and $5 per visit
    nonmembers: $10 per visit

    Write and solve an equation to find the number of visits each year that will make a member's total cost less than a nonmember's total cost. *(Section 4.5)*

Solve each inequality. Graph each solution on a number line. If there is no solution, write *no solution*. *(Section 4.6)*

16. $-2x + 3 < 2x + 3$

17. $6y \geq 7y + 4$

18. $2(3y + 1) \leq 6y$

**SPIRAL REVIEW    Chapters 1–4**

Find the slope of the line through each pair of points.

1. $(2, -3)$ and $(6, 17)$

2. $(7, 1)$ and $(7, -8)$

**For Exercises 3–8, solve each equation.**

**3.** $x + 22 = -7$

**4.** $\frac{3}{5}y = 30$

**5.** $12 - \frac{b}{4} = 2$

**6.** $-3 - 2a = 3$

**7.** $5p - (p + 2) = 18$

**8.** $3 - 7u = 2(3u - 5)$

**9.** Tell whether the data in the table show direct variation. If they do, give the constant of variation and write an equation that describes the situation.

| Pounds of bananas sold | Cost |
|---|---|
| 2 | $0.78 |
| 5 | $1.95 |
| 15 | $5.85 |

**Write an equation of the line with the given slope and through the given point.**

**10.** slope = 0; (–4, –1)

**11.** slope = $\frac{1}{2}$; (2, 12)

**12.** undefined slope; (3, 3)

**13.** slope = –4; (2, 3)

**Graph each inequality on a number line.**

**14.** $x > 0$

**15.** $a \geq -3$

**16.** $n \leq -1$

**Graph each equation.**

**17.** $y = 2x - 1$

**18.** $y = \frac{2}{3}x + 1$

**19.** $y = \frac{1}{3}x - 2$

**For Exercises 20–22, simplify each expression.**

**20.** $0 - 3(6 + 5)$

**21.** $\frac{15}{3} - 5$

**22.** $2(x + 1) - 9(x - 3)$

**23.** Find the mean, the median, and the modes(s) of the data below.

Student's distance from school (mi): 3, 4, 2, 2, 5, 4, 6, 5, 3, 3, 8, 6

**24.** What is the least common multiple of 3, 6, and 10?

**25.** Marietta made and sold jewelry at a craft fair. She paid $16.00 for the use of a booth at the fair and spent $2.75 for the materials for each piece of jewelry. She sold each piece of jewelry for $5.00. Write an equation for her profit as a function of the number of pieces of jewelry she sold. Describe the domain and range of this function.

**Study Guide, ALGEBRA 1: EXPLORATIONS AND APPLICATIONS**

# Section 5.1 Ratios and Proportions

**GOAL**

**Learn how to...**
- use ratios to compare quantities
- solve proportions

**So you can...**
- estimate quantities that are difficult to count

## Application

You need to read a 180-page book for history class. You have read 35 pages in two days. You can use ratios and proportions to find out how long it will take you to finish the book.

### Terms to Know / Example / Illustration

| Terms to Know | Example / Illustration |
|---|---|
| **Ratio (p. 192)** — a quotient that compares two quantities | There are 13 girls and 11 boys in one class. The ratio of girls to boys can be written three ways: $\frac{13}{11}$   13:11   13 to 11 |
| **Proportion (p. 193)** — an equation that shows two ratios to be equal (Proportions can be written three ways: $\frac{a}{b} = \frac{c}{d}$, $a:b = c:d$, and $a$ to $b = c$ to $d$.) | $\frac{3}{5} = \frac{x}{20}$ |
| **Means (p. 193)** — in the proportion $\frac{a}{b} = \frac{c}{d}$, the values $b$ and $c$ | In the proportion above, 5 and $x$ are the means. |
| **Extremes (p. 193)** — in the proportion $\frac{a}{b} = \frac{c}{d}$, the values $a$ and $d$ | In the proportion above, 3 and 20 are the extremes. |

# UNDERSTANDING THE MAIN IDEAS

## Ratios

When you write a ratio, the two quantities must be measured in the same units.

### Example 1

Express as a ratio: 7 girls to 6 boys.

**Solution**

$\frac{7}{6}$ or 7:6 or 7 to 6

### Example 2

What is the ratio of 7 in. to 2 ft?

**Solution**

Since 1 ft = 12 in., change 2 ft to 24 in.

So the ratio is $\frac{7}{24}$ or 7:24.

**Find the ratio of the first quantity to the second. Write your answer as a fraction in lowest terms.**

1. 6 to 8
2. 12 to 36
3. 21 to 28
4. 6 million motorcycles to 120 million cars
5. 4 million people in Alabama to 17 million people in Texas
6. 10 in. to 2 ft
7. speed of cheetah (70 mi/h) to speed of giraffe (32 mi/h)

## Proportions

In a proportion, the product of the means is equal to the product of the extremes. This is called the *means-extremes property*.

if $\frac{a}{b} = \frac{c}{d}$, then $ad = bc$ → if $\frac{3}{5} = \frac{6}{10}$, then $3(10) = 5(6)$

You can use the *means-extremes property* to solve proportions.

**Study Guide,** ALGEBRA 1: EXPLORATIONS AND APPLICATIONS
Copyright © McDougal Littell Inc. All rights reserved.

## Example 3

You need to read a 180-page book. It takes you 2 days to read 35 pages. How long will it take you to read the whole book?

### Solution

Use a proportion: 2 days is to 35 pages as $x$ days is to 180 pages.

$$\frac{2 \text{ days}}{35 \text{ pages}} = \frac{x \text{ days}}{180 \text{ pages}} \rightarrow \frac{2}{35} = \frac{x}{180}$$

$$35x = 2(180) \quad \leftarrow \text{Use the means-extremes property.}$$

$$35x = 360$$

$$\frac{35x}{35} = \frac{360}{35}$$

$$x = \frac{72}{7}$$

$$x \approx 10.3$$

It will take you about 10 days to read the book.

## Example 4

In major league baseball, there are approximately 2 left-handed pitchers for every 5 right-handed pitchers. In a group of 50 pitchers, how many left-handers would you expect to find?

### Solution

Since there are about 2 left-handers for every 5 right-handers, then about 2 out of every 7 pitchers are left-handed.

$$\frac{2 \text{ left-handers}}{7 \text{ pitchers}} = \frac{x \text{ left-handers}}{50 \text{ pitchers}} \rightarrow \frac{2}{7} = \frac{x}{50}$$

$$7x = 2(50)$$

$$7x = 100$$

$$\frac{7x}{7} = \frac{100}{7}$$

$$x = \frac{100}{7}$$

$$x \approx 14.3$$

You should expect about 14 left-handed pitchers.

8. Identify the means and the extremes in the proportion $\dfrac{5}{x} = \dfrac{7}{9}$.

**For Exercises 9–15, solve each proportion.**

9. $\dfrac{3}{x} = \dfrac{5}{9}$

10. $\dfrac{y}{12} = \dfrac{5}{7}$

11. $\dfrac{2.5}{1.7} = \dfrac{3.1}{d}$

12. $9:w = 6:13$

13. 13 to $t$ = 39 to 12

14. $\dfrac{x+1}{x} = \dfrac{14}{12}$

15. $\dfrac{2x-1}{3} = \dfrac{x+1}{2}$

16. Biologists catch, tag, and release 49 catfish in a pond. Several days later they catch 31 catfish, of which 11 have tags. About how many catfish are in the pond?

17. Two gears are shown in the figure at the right. The gear ratio is 3:2. This means that every time gear A makes 3 complete revolutions, gear B makes exactly 2 revolutions. If gear A revolves 10,000 times, how many revolutions does gear B make?

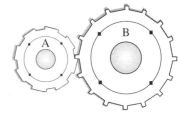

18. Melika says that the equation $\dfrac{2}{3} = 5 + \dfrac{1}{x}$ is not a proportion. Explain why she is correct.

19. The ratio of ninth graders to tenth graders in a school soccer program is 3 to 5. There are 48 players in all. How many ninth graders are there in the program?

## Spiral Review

**Find an equation of the line through the given points.** *(Section 3.5)*

20. (3, 5) and (1, –2)

21. (–4, –1) and (–1, 4)

22. (2, –6) and (2, 3)

23. (–1, 5) and (4, 5)

**Solve each equation. Round answers to the nearest hundredth.** *(Section 4.4)*

24. $1.34x - 5.26 = 7.83$

25. $\dfrac{8.73(n + 12.62)}{15.66} = -14.35$

26. $6.85 - 19.72b = -22.06$

27. $2.37r + 8.74r = 54.97$

# Section 5.2 Scale Measurements

**GOAL**

**Learn how to...**
- use scales in drawings and on graphs
- use scale factors to compare sizes of drawings

**So you can...**
- model large objects with scaled drawings
- find large dimensions
- compare maps of different sizes

## Application

Suppose you are driving from Maryland to Florida. On a large map of the United States, the scale is 1 in. = 100 mi, and the map distance for your trip is about 9 in. When you reach the Florida state line, you switch to a map where the scale is 1 in. = 20 mi. The distance from Jacksonville to Orlando on the map is about 6 in. How can you compare the actual distances?

| Terms to Know | Example / Illustration |
|---|---|
| **Scale (p. 198)**<br>a comparative measurement indicating the relationship between an actual measurement and the length representing that measurement on a drawing | 1 in. = 20 mi<br><br>1 cm = 100 km<br><br>$\frac{1}{4}$ in. = 10 ft |
| **Scale factor (p. 199)**<br>a ratio used to show how the scale of a drawing, photograph, or map has changed | In the Application, the map of Florida is 5 times the size of Florida on the map of the United States. So, the scale factor of the two maps is 5:1. |

Study Guide, ALGEBRA 1: EXPLORATIONS AND APPLICATIONS

# UNDERSTANDING THE MAIN IDEAS

## Scales on drawings

The *scale factor* of the two drawings shown below is given by the ratio

$$\frac{\text{wingspan of new model}}{\text{wingspan of old model}} = \frac{0.56}{1.25} \approx 0.45, \text{ or approximately } 45\%.$$

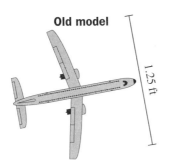
Old model

New model

### Example 1

The propeller on Rico's $\frac{1}{32}$-scale model of an airplane is 4.5 in. long. How long is the actual airplane propeller? (*Hint*: A $\frac{1}{32}$-scale means 1 in. = 32 in.)

### ■ Solution ■

Let $p$ = the length of the actual propeller. Use the scale to write a proportion.

$\frac{1}{32} = \frac{4.5}{p}$ ← The ratios are shown in the form $\frac{\text{length of model propeller}}{\text{length of actual propeller}}$.

$1(p) = 32(4.5)$ ← Use the means-extreme property.

$p = 144$

The length of the actual propeller is 144 in.

### Example 2

Kaitlin has a picture that is 8 in. wide by 10 in. long. She wants to use a photocopier to reduce the picture so that it is just 4 in. long.

a. What scale factor should she use to make the reduction?

b. How wide will the reduced picture be?

> **Solution**
>
> **a.** The scale factor is $\frac{4}{10} = 0.4$, or 40%. This means the photocopier should be set for a 40% reduction.
>
> **b.** The width of the reduced picture will be 0.4(8 in.) = 3.2 in.

**In Exercises 1–3, the dimensions of a photograph are given. Badri needs to reduce each of these photographs so that the longer side is 3 in. long. What scale factor should he use, and what will be the length of the shorter side?**

1. 4 in. by 6 in.
2. 5 in. by 9 in.
3. 11 in. by 15 in.

**The scale on a map of New Jersey says that 3 in. represents 8 mi. For Exercises 4 and 5, find each actual distance to the nearest tenth of a mile.**

4. Newark International Airport to Livingston, 4 in. on the map
5. Newark International Airport to Princeton, 14 in. on the map

6. On the same map of New Jersey discussed above, about how many inches would separate two towns that are actually 28 mi apart?

7. When the scale on a map says that 3 in. = 8 mi, how many actual *inches* of distance are represented by each inch on that map? (*Note*: Recall that 1 mi = 5280 ft.)

8. **Writing** Look at the two world bicycle production graphs shown below. Explain how the choice of vertical scale changes the way a reader might interpret the information.

| World Bicycle Production ||
|---|---|
| Year | Number of bicycles (in millions) |
| 1960 | 20 |
| 1970 | 36 |
| 1980 | 62 |
| 1990 | 95 |

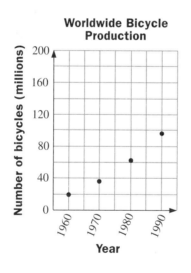
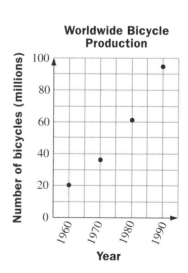

9. **Open-ended Problem** Draw two pictures where the scale factor is 2:3.

## Spiral Review

**Solve each inequality. Graph the solution on a number line.** *(Section 4.6)*

**10.** $2x - 5 > 7$

**11.** $2(x + 5) < 10$

**12.** $4 - 3x \leq 21$

**13.** $\frac{2}{3}x \geq -8$

**14.** $5 + 3x \leq 2\left(\frac{3}{4}x + 2\right)$

**15.** $15x \geq \frac{5}{3}(1 + 12x)$

# Section 5.3 Working with Similarity

**GOAL**

Learn how to . . .
- recognize and create similar figures

So you can . . .
- measure large distances in real-life situations

## Application

You can use similar triangles to estimate the height of tall buildings, trees, and other objects.

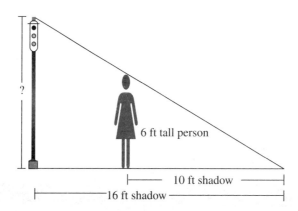

## Terms to Know

**Similar (p. 206)**
a property of two figures for which corresponding angles have equal measure and the ratios of the lengths of corresponding sides are equal (The symbol for similarity is ~.)

## Example / Illustration

The figures below are similar.

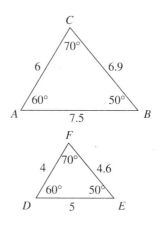

$\angle A = \angle D = 60°;$
$\angle B = \angle E = 50°;$
$\angle C = \angle F = 70°;$
$\dfrac{AB}{DE} = \dfrac{BC}{EF} = \dfrac{AC}{DF} = \dfrac{3}{2}$

Study Guide, ALGEBRA 1: EXPLORATIONS AND APPLICATIONS
Copyright © McDougal Littell Inc. All rights reserved.

## UNDERSTANDING THE MAIN IDEAS

Similar figures have the same *shape,* but not always the same *size*. When naming similar figures, always name the corresponding angles in the same order. For the triangles shown on page 108, the similarity is written △ABC ~ △DEF.

*Note:* Two triangles are similar when *two* pairs of corresponding angles are similar. This is true because the third pair of angles must also have equal measure.

### Example 1

In the figure at the right, △LMN is similar to △RST. Use similar figures to find RS.

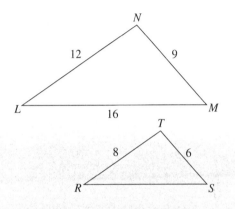

### Solution

Use a proportion. The ratios of the lengths of corresponding sides are equal.

$$\frac{LN}{RT} = \frac{LM}{RS} \rightarrow \frac{12}{8} = \frac{16}{RS}$$

$$12 \cdot RS = 128 \quad \leftarrow \text{Use the means-extremes property.}$$

$$\frac{12 \cdot RS}{12} = \frac{128}{12}$$

$$RS = \frac{32}{3}, \text{ or } 10\frac{2}{3}$$

**For Exercises 1–4, use the figure at the right.
Triangle *FTP* is similar to triangle *GMW*.**

1. What can you determine about ∠P and ∠W?
2. What side corresponds to $\overline{FT}$?
3. What side corresponds to $\overline{WM}$?
4. What other two ratios are equal to $\frac{FT}{GM}$?

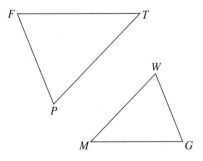

**For Exercises 5–7, use the figure at the right.
Quadrilateral *MNOP* is similar to quadrilateral *QRST*.**

5. What is the measure of $\angle R$?
6. What is the length of $\overline{RQ}$?
7. What is the length of $\overline{PO}$?

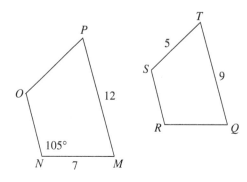

### Example 2

Use similar triangles to determine the height of the tree. (Notice that the two triangles share one angle and that both have a second angle that is a right angle.)

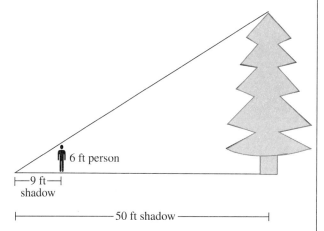

### ■ Solution ■

The situation involves the two similar triangles shown at the right. Use a proportion. Let $x$ = the height of the tree.

$$\frac{9}{50} = \frac{6}{x}$$

$$9x = 300$$

$$x = \frac{100}{3}, \text{ or } 33\frac{1}{3}$$

The tree is about $33\frac{1}{3}$ ft tall.

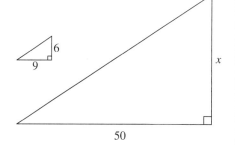

**Study Guide,** ALGEBRA 1: EXPLORATIONS AND APPLICATIONS
Copyright © McDougal Littell Inc. All rights reserved.

8. Naomi sighted from the ground past Samantha's head to the tip of a tree in the city park. They know that Samantha is 1.7 m tall and they measured the other distances shown in the figure. How tall is the tree?

9. **Open-ended Problem** Draw two similar triangles where the ratio of corresponding sides is 1:2.

10. **Mathematics Journal** Describe the connections between ratios, proportions, similar figures, and solving an equation.

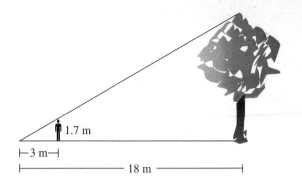

## Spiral Review

**Express each rate in the given unit. Write each answer as a unit rate.**
*(Section 3.1)*

11. 60 miles per hour = _?_ feet per second

12. 16.2 kilometers per liter = _?_ centimeters per milliliter

13. 1200 parts per day = _?_ parts per week

**Tell whether the data show direct variation. If they do, give the constant of variation and write an equation.** *(Section 3.2)*

14.

| Height (in.) | Shoe size |
|---|---|
| 64 | 5 |
| 67 | 7 |
| 69 | 9 |
| 72 | 10 |
| 75 | 14 |

15.

| Vehicle weight (lb) | Registration fee (dollars) |
|---|---|
| 2200 | 176 |
| 2350 | 188 |
| 2480 | 198 |
| 2600 | 208 |
| 2880 | 230 |

# Section 5.4

## Perimeters and Areas of Similar Figures

**GOAL**

**Learn how to . . .**
- find the ratios of the perimeters and areas of similar figures

**So you can . . .**
- use scale drawings to find the perimeters and areas of houses, for example
- create accurate pictographs to represent statistics

### Application

Suppose the floor plan of a house has a scale where 1 in. = 10 ft. The perimeter of the house on the floor plan is 14 in. and its area is 12 in.$^2$. What are the perimeter and area of the actual house?

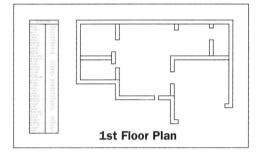
1st Floor Plan

## UNDERSTANDING THE MAIN IDEAS

The two figures shown below are similar. The *scale factor*, A to B, is 1:3.

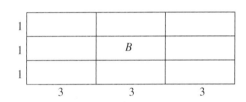

Since the scale factor is 1:3, the ratio of the lengths of corresponding sides of the two similar figures is 1:3. Likewise, the ratio of the perimeters of the two figures is 1:3.

> Figure A perimeter: 3 + 3 + 1 + 1 = 8 units
> Figure B perimeter: 9 + 9 + 3 + 3 = 24 units

The ratio of the perimeters (A to B) is 8:24, or 1:3, the same as the scale factor.

The ratio of the areas of two similar figures is also related to the scale factor. Since the scale factor for the two figures above is 1:3, the ratio of the areas will be $1^2:3^2$, or 1:9.

> Figure A area: $1 \times 3 = 3$ square units
> Figure B area: $3 \times 9 = 27$ square units

The ratio of the areas (A to B) is 3:27, or 1:9.

In general, if the ratio of the lengths of corresponding sides of two similar figures is $a:b$, then the ratio of the perimeters is also $a:b$ and the ratio of the areas is $a^2:b^2$.

### Example 1

Quadrilaterals *ABCD* and *RSTU* are similar.

a. Find the scale factor of the similar figures.

b. Find the perimeter and area of both quadrilaterals.

c. Find the ratio of the perimeters and the ratio of the areas.

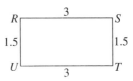

### Solution

a. The scale factor (*ABCD* to *RSTU*) is 1:1.5, or 2:3.

b. perimeter of *ABCD*: $1 + 1 + 2 + 2 = 6$ cm; area of *ABCD*: $1 \times 2 = 2$ cm$^2$
perimeter of *RSTU*: $1.5 + 1.5 + 3 + 3 = 9$ cm; area of *RSTU*: $1.5 \times 3 = 4.5$ cm$^2$

c. $\dfrac{\text{perimeter of } ABCD}{\text{perimeter of } RSTU} = \dfrac{6}{9} = \dfrac{2}{3}$  ← same ratio as the scale factor

$\dfrac{\text{area of } ABCD}{\text{area of } RSTU} = \dfrac{2}{4.5} = \dfrac{4}{9}$  ← the square of the scale factor

---

**Use similar quadrilaterals *PQRS* and *WXYZ* shown at the right.**

1. Find the scale factor from quadrilateral *PQRS* to quadrilateral *WXYZ*.
2. Find the ratio of the perimeters.
3. Find the ratio of the areas.

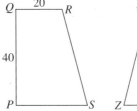
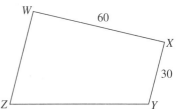

### Example 2

Pentagons *DEFGH* and *JKLMN* are similar.

a. What is the perimeter of pentagon *JKLMN*?

b. What is the area of pentagon *DEFGH*?

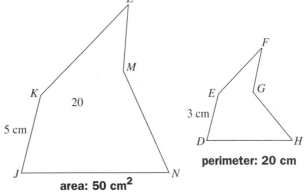

**Study Guide,** ALGEBRA 1: EXPLORATIONS AND APPLICATIONS

■ Solution ■

a. The ratio of the perimeters of two similar figures is the same as the scale factor. The scale factor (DEFGH to JKLMN) is 3:5.

$$\frac{\text{perimeter of } DEFGH}{\text{perimeter of } JKLMN} = \frac{3}{5}$$

$\frac{20}{x} = \frac{3}{5}$  ← Let $x$ = the perimeter of pentagon JKLMN.

$3x = 100$  ← Use the means-extremes property.

$\frac{3x}{3} = \frac{100}{3}$

$x = \frac{100}{3}$

The perimeter of pentagon JKLMN is $\frac{100}{3}$ cm, or about 33.3 cm.

b. The ratio of the areas of two similar figures is the same as the square of the scale factor. Again, the scale factor (DEFGH to JKLMN) is 3:5.

$$\frac{\text{area of } DEFGH}{\text{area of } JKLMN} = \frac{3^2}{5^2}$$

$\frac{x}{50} = \frac{9}{25}$  ← Let $x$ = the area of pentagon DEFGH.

$25x = 450$  ← Use the means-extremes property.

$\frac{25x}{25} = \frac{450}{25}$

$x = 18$

The area of pentagon DEFGH is 18 cm$^2$.

**For Exercises 4 and 5, use similar pentagons BCDEF and JKLMN shown at the right.**

4. Find the area of pentagon BCDEF.

5. Find the perimeter of pentagon JKLMN.

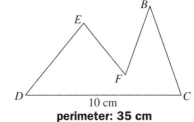
perimeter: 35 cm

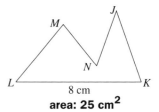
area: 25 cm$^2$

6. Two similar figures have areas of 81 in.$^2$ and 144 in.$^2$. If the perimeter of the smaller figure is 60 in., what is the perimeter of the larger figure?

7. The perimeters of two similar figures are 24 cm and 40 cm. find the area of the smaller figure if the area of the larger figure is 150 cm$^2$.

## Spiral Review

**Solve each equation.** *(Sections 4.2 and 4.3)*

8. $3(x - 7) = 27$

9. $\frac{2}{3}x + 5 = 17$

10. $-3(2 - x) = 21$

11. $\frac{2z - 3}{2} = 3.8$

12. $2.4x - 5.8 = -5.8$

13. $\frac{3}{x - 5} = 2$

# Section 5.5

# Exploring Probability

**GOAL**

**Learn how to . . .**

- calculate experimental and theoretical probability

**So you can . . .**

- predict the outcome of events

## Application

We can never be certain about what the weather will be tomorrow, but we might know, for example, that under the same conditions in the past, it rained the next day 60% of the time. What exactly does this mean?

### Terms to Know / Example / Illustration

| Terms to Know | Example / Illustration |
|---|---|
| **Probability (p. 218)** a ratio that measures how likely something is to happen (Probability is sometimes expressed as a percent.) | Chantrea has made 70% of her free throws during this basketball season. The probability that she will make her next free throw is $\frac{7}{10}$, or 0.7, or 70%. |
| **Outcome (p. 218)** the result of an experiment, such as rolling a die | There are six outcomes when a standard die is rolled: 1, 2, 3, 4, 5, 6. |
| **Event (p. 218)** a set of one or more outcomes of an experiment | Getting an even number is one event for the roll of a die. (Notice that this event is made up of three of the six possible outcomes). |
| **Experimental probability (p. 219)** the ratio of the number of times an event actually occurs to the total number of times the experiment is done<br><br>Experimental probability = $\frac{\text{Number of successes}}{\text{Number of tries}}$ | A coin is tossed 50 times and heads is the result 28 times. The experimental probability of getting heads for this coin is $\frac{28}{50}$, or 0.56 (or 56%). |

Study Guide, ALGEBRA 1: EXPLORATIONS AND APPLICATIONS
Copyright © McDougal Littell Inc. All rights reserved.

| **Theoretical probability** (p. 220) the ratio of the number of outcomes that make the event happen to the total number of outcomes | Since you are equally likely to get heads as tails when tossing a coin, the theoretical probability of tossing heads is $\frac{1}{2}$, or 0.5. |
|---|---|

## UNDERSTANDING THE MAIN IDEAS

Probability is a measure of how likely a certain event will happen. It can help you to understand sports, politics, science, and other aspects of your life.

### Example 1

During his professional career from 1914 to 1935, Babe Ruth had 2873 hits in 8399 at-bats. What was the probability that he would get a hit in a given at-bat?

### Solution

The 2873 hits were his "successes" and the 8399 at-bats were his "attempts."

$$\text{probability of a hit} = \frac{2873}{8399} \approx 0.342, \text{ or about } 34.2\%$$

(*Note:* In baseball, this probability is called the "batting average," and the decimal form is used. Ruth had a career batting average of .342.)

### Example 2

What is the theoretical probability that you will roll a number greater than 4 on one roll of a standard die?

### Solution

The possible outcomes are 1, 2, 3, 4, 5, and 6. Only the 5 and 6 are successes. The theoretical probability for each of the numbers on a die is $\frac{1}{6}$. So the theoretical probability of rolling a number larger than 4 is $\frac{1}{6} + \frac{1}{6} = \frac{2}{6} \approx 0.333$, or about 33.3%.

**Study Guide,** ALGEBRA 1: EXPLORATIONS AND APPLICATIONS
Copyright © McDougal Littell Inc. All rights reserved.

**Toss a coin 20 times and record whether you get heads or tails.**

1. What is your experimental probability of getting heads?
2. What is your experimental probability of getting tails?
3. What is the sum of your answers to Exercises 1 and 2?
4. **Writing** Explain why your answer to Exercise 3 makes sense.

**For Exercises 5–9, roll a die 36 times and record how many times you get each of the six possible outcomes.**

5. Make a histogram of the data you collected, with the numbers 1 through 6 along the horizontal axis.
6. What is the experimental probability of getting each possible outcome?
7. What is the theoretical probability of getting each possible outcome?
8. Suppose the results of your 36 rolls exactly matched the theoretical probability. What would your histogram for Exercise 5 look like?
9. **Writing** Would you be suprised if your experimental probabilities were exactly equal to the theoretical probability for each possible outcome? Explain.
10. In 1987, Roger Clemens won 20 games and lost 9 for the Boston Red Sox, for 29 "decisions" in all. For the games in which he got the "decision," what was the probability that he won a given game during that season?
11. A pitcher won $W$ games and lost $L$ games during a season. Write **(a)** an expression for the total number of decisions for the pitcher that season, and **(b)** an expression for the experimental probability that the pitcher won a game in which he got the decision.

**TECHNOLOGY** Many calculators can generate random numbers. On a graphing calculator, press the MATH key, select PRB, and then choose Rand. Pressing ENTER repeatedly, the calculator generates random decimals between 0 and 1.

12. Generate 20 random numbers on a calculator and record each first digit.
13. What is your experimental probability for getting the digit 3?
14. What is the theoretical probability of getting the digit 3?

### Spiral Review

**Graph each point in a coordinate plane. Name the quadrant (if any) in which the point lies.** *(Section 2.4)*

15. $A(2, -5)$
16. $B(-3, 0)$
17. $C(-2, -1)$

**Solve each equation. Check your solutions.** *(Section 4.4)*

18. $\dfrac{b}{48} = \dfrac{7}{12}$
19. $\dfrac{x}{4} + \dfrac{3}{4} = 6$
20. $2(n + 3) = 3(n + 1)$

# Section 5.6 Geometric Probability

**GOAL**

**Learn how to . . .**
- find probabilities based on area

**So you can . . .**
- predict events

## Application

You are playing a board game where you throw a chip onto a board that has a white region and a black region. You get more points for landing on the black region. You can use geometric probability to determine your chances of landing on this black region.

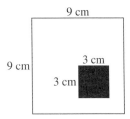

### Terms to Know

**Geometric probability** (p. 224)
probability based on length or area

### Example / Illustration

The probability of landing in the black area of the game board shown in the Application above is

$$\frac{\text{black area}}{\text{total area}} = \frac{9}{81} = \frac{1}{9}.$$

## UNDERSTANDING THE MAIN IDEAS

In all situations where you can apply geometric probability, it is important to compare the area of special interest to the *total area*, such as the winning area on a game board or target. The geometric probability is given by the ratio below.

$$\frac{\text{area of special interest}}{\text{total area}}$$

### Example 1

Find the probability of a randomly-thrown dart hitting the shaded area of the target shown at the right. Assume that you are equally likely to hit each point on the target. The small squares are all the same size.

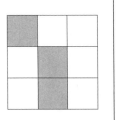

**Study Guide,** ALGEBRA 1: EXPLORATIONS AND APPLICATIONS

### Solution

There are 3 shaded squares and 9 squares in all. Let $x$ = the area of one small square.

$$\frac{\text{area of shaded squares}}{\text{total area}} = \frac{3x}{9x} = \frac{1}{3}$$

So, the probability of the dart hitting the shaded square is $\frac{1}{3}$, or about 33.3%.

## Example 2

The game board at the right is in the shape of an equilateral triangle. The shaded triangular region is also equilateral. What is the probability of hitting the shaded region? Assume that you are equally likely to hit each point on the target.

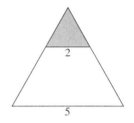

### Solution

All equilateral triangles are similar (the angles measure 60°). The scale factor here is 2:5, so the ratio of the area of the shaded region to the whole region is $2^2:5^2$ or 4:25. Therefore, the probability is $\frac{4}{25} = 0.16$, or 16%.

## Example 3

Find the probability that a randomly-thrown dart will land in one of the shaded regions on the target shown at the right.

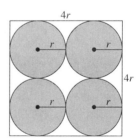

### Solution

Since the area of each of the four circles is $\pi r^2$, the total area of the shaded regions is $4\pi r^2$. The length of each side of the square is $4r$, so its total area is $(4r)(4r) = 16r^2$.

So the probability of landing in a shaded region is $\frac{4\pi r^2}{16r^2} = \frac{\pi}{4} \approx 0.785$, or about 78.5%.

For Exercises 1–6, find the probability that a randomly-thrown dart hits the shaded area of each target. The figures are squares, equilateral triangles, and circles.

1.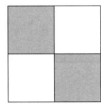

2.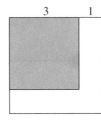

3.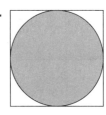

4.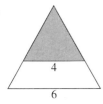

5.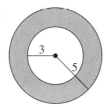

6.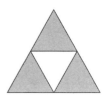

7. **Open-ended Problem** Draw a game board design where the probability of hitting the shaded area is $\frac{5}{8}$.

8. **Open-ended Problem** Draw a game board design where the probability of hitting the shaded area is $\frac{1}{6}$.

9. A Coast Guard helicopter is searching for a boat within a 1000 square mile region. On this particular day the pilots can see for a radius of 5 mi. If they are searching at random, what is the probability that they will see the boat at any given time?

10. **Mathematics Journal** Describe some of the advantages of representing probability situations geometrically.

## Spiral Review

**Evaluate each variable expression for $a = -2$ and $b = \frac{1}{3}$.** *(Sections 1.2, 1.4, 1.5)*

11. $a - b$

12. $a^2 - b^2$

13. $-a^2 + 3b$

14. $2b(a - a)$

15. $\left(\frac{a+b}{4}\right)(a - 10)$

16. $\frac{-a + 3b}{b + 2(b - a)}$

# Chapter 5 Review

**CHAPTER CHECK-UP**

Complete these exercises for a review of Chapter 5. If you have difficulty with a particular problem, review the indicated section.

1. Express the ratio 10 teachers to 250 students in reduced form. *(Section 5.1)*

2. Solve the proportion $\dfrac{2.5}{1.8} = \dfrac{3.1}{k}$. *(Section 5.1)*

3. An 8 in. by 12 in. photograph is reduced so that its longer side is 9 in. long. How long is the shorter side? *(Section 5.2)*

4. A picture of a cat shows his head to be 5 in. wide. A reduced version of the same picture shows his head to be 2 in. wide. What is the scale factor? *(Section 5.2)*

5. **Writing** The two graphs below show the same data on T-shirt sales as a function of temperature. Compare the scales of the two graphs. *(Section 5.3)*

**T-Shirt Sales at a Store**

**T-Shirt Sales at a Store**

**Use the figure at the right.** *(Section 5.3)*

6. Explain how you know that the two triangles are similar.
7. What is the ratio of $x$ to $y$?

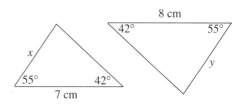

**Use the similar figures at the right.** *(Section 5.4)*

8. What is the scale factor of figure $A$ to figure $B$?
9. What is the perimeter of figure $B$?
10. What is the area of figure $A$?

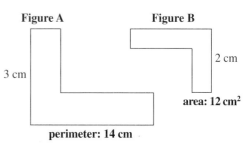

**11. Open-ended Problem** Describe an event with a probability of 1 and another event with a probability of about 0.5. *(Section 5.5)*

**12.** Jake made 8 of his 12 free throws in the last two games. During that time, what was the the experimental probability that he would make a free throw? *(Section 5.5)*

**The following figures are either squares, circles, or equilateral triangles. Find the probability that a randomly-thrown dart hits the shaded area of each figure.** *(Section 5.6)*

**13.** 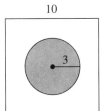  **14.** 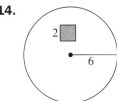  **15.**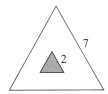

## SPIRAL REVIEW  Chapters 1–5

The table at the right shows the growth in the number of answering machines sold in the world.

1. In planning a graph of the data with the years along the horizontal axis, what vertical scale would you use?
2. Graph the data. Draw a line between the points for 1986 and 1992. What is the slope of the line?
3. Use the line you drew for Exercise 2 to predict the sales in 1996.

| Year | Number of machines sold |
|---|---|
| 1986 | 6,450,000 |
| 1988 | 11,100,000 |
| 1990 | 13,800,000 |
| 1992 | 15,482,000 |

The per-game averages of the highest scoring NBA players during the 11 seasons from 1980 to 1990 are given below. Use this information for Exercises 4–6.

33.1, 30.7, 32.3, 28.4, 30.6, 32.9, 30.3, 36.9, 35.1, 32.5, 33.6

**4.** Find the mean.   **5.** Find the median.   **6.** Find the mode.

**7.** What is the probability that a randomly-thrown dart will hit the shaded area in the figure at the right? The triangles are all equilateral.

**Evaluate each expression for $a = -2$ and $b = 0.5$.**

**8.** $a - b$   **9.** $(a-b)(a+b)$   **10.** $-a^2 + b$

**Solve each equation or inequality.**

**11.** $1.3x - 2.5 = 3.4$   **12.** $3(2x - 7) = 27$   **13.** $5 - 2x < 15$

**14.** $\dfrac{x-4}{3} = \dfrac{x}{6}$   **15.** $\dfrac{x}{3} - 5 > 7$   **16.** $4 + 3x \le 0$

# Section 6.1 The Pythagorean Theorem

**GOAL**

Learn how to . . .
- identify right triangles using the Pythagorean theorem

So you can . . .
- solve real-world problems

*Application*

Students building a set for the school play need to cut wood for the frame of a backdrop and for a diagonal brace. The vertical, horizontal, and diagonal lengths are related by the Pythagorean theorem.

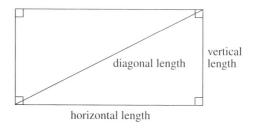

### Terms to Know | Example / Illustration

**Pythagorean theorem (p. 239)**
the theorem which states that "in a right triangle, the square of the length of the hypotenuse is equal to the sum of the squares of the lengths of the legs"

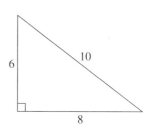

$6^2 + 8^2 = 36 + 64 = 100$
$10^2 = 100$
So, $6^2 + 8^2 = 10^2$.

**Right triangle (p. 239)**
a triangle with one right or 90° angle

**Leg (p. 239)**
the two sides other than the hypotenuse in a right triangle (These two sides form the right angle.)

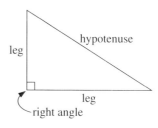

**Hypotenuse (p. 239)**
the long side opposite the right angle in a right triangle

**Perfect square (p. 240)**
a number that is the square of another number

0, 1, 4, 9, 16, 25, ...

| Terms to Know | Example / Illustration |
|---|---|
| **Square root (p. 240)**<br>a number whose square is a particular number | The positive square root of 9 is 3, and the negative square root of 9 is $-3$. |
| **Radical sign (p. 240)**<br>the symbol $\sqrt{\phantom{x}}$ that indicates the positive square root of a number | $\sqrt{36} = 6$<br>$\sqrt{\dfrac{1}{4}} = \dfrac{1}{2}$ |
| **Converse of the Pythagorean theorem (p. 241)**<br>the theorem which states that "if the lengths of the sides of a triangle have the relationship $a^2 + b^2 = c^2$, then the triangle is a right triangle" | 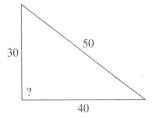<br>$30^2 + 40^2 = 900 + 1600 = 2500 = 50^2$<br>The triangle is a right triangle. |

## UNDERSTANDING THE MAIN IDEAS

### *Radical sign*

The radical sign $\sqrt{\phantom{x}}$ is used to refer to the positive square root of a number.

---

**Example 1**

Find each square root.

    **a.** $\sqrt{25}$                           **b.** $-\sqrt{100}$

■ **Solution** ■

   **a.** Rewrite the number inside the radical sign as a square.

      $\sqrt{25} = \sqrt{5 \cdot 5} = 5$

   **b.** Notice that the negative sign is outside the radical sign.

      $-\sqrt{100} = -\left(\sqrt{10 \cdot 10}\right) = -(10) = -10$

**For Exercises 1–6, find each square root.**

1. $\sqrt{81}$
2. $\sqrt{1}$
3. $-\sqrt{10000}$
4. $-\sqrt{169}$
5. $\sqrt{441}$
6. $-\sqrt{900}$

7. The sum of a number and its square root is 30. What is the number?

## The Pythagorean theorem

In a right triangle, the square of the length of the hypotenuse is equal to the sum of the squares of the lengths of the legs.

$c^2 = a^2 + b^2$

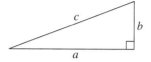

### Example 2

Use the Pythagorean theorem to find the missing side length for each right triangle.

a.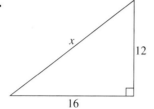

b.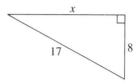

■ **Solution** ■

a. The missing length is the hypotenuse.

$x^2 = 12^2 + 16^2$
$\phantom{x^2} = 144 + 256$
$\phantom{x^2} = 400$
$x = \sqrt{400}$
$\phantom{x} = 20$

b. The missing length is one of the legs.

$8^2 + x^2 = 17^2$
$64 + x^2 = 289$
$x^2 = 225$
$x = \sqrt{225}$
$\phantom{x} = 15$

For Exercises 8–10, find the unknown length for each right triangle.

8.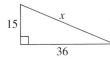
9.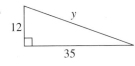
10.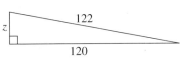

11. A firefighter's ladder, 61 ft long, rests against a tall building. If the bottom of the ladder is 11 ft from the building, how high up the wall does the ladder reach?

12. One rectangle is 48 ft long and 14 ft wide. Another rectangle is 40 ft long and 30 ft wide. Which rectangle has a longer diagonal?

## *Converse of the Pythagorean theorem*

If the lengths of the sides of a triangle have the relationship $a^2 + b^2 = c^2$, then the triangle is a right triangle.

### Example 3

Use the converse of the Pythagorean theorem to decide whether the given lengths can be the sides of a right triangle.

    **a.** 16 cm, 30 cm, 34 cm     **b.** 2 ft, 3 ft, 4 ft

### ■ Solution ■

**a.** The greatest length is 34, so compare $34^2$ with $16^2 + 30^2$.

$$34^2 = 1156$$
$$16^2 + 30^2 = 256 + 900 = 1156$$

Since $34^2 = 16^2 + 30^2$, the triangle is a right triangle.

**b.** The greatest length is 4, so compare $4^2$ with $2^2 + 3^2$.

$$4^2 = 16$$
$$2^2 + 3^2 = 4 + 9 = 13$$

Since $4^2 \neq 2^2 + 3^2$, the triangle is not a right triangle.

For Exercises 13–16, tell whether the given lengths can be the sides of a right triangle.

13. 3.5 cm, 4.7 cm, 6.0 cm
14. 24 m, 45 m, 51 m
15. 9 m, 40 m, 41 m
16. 3 in., 4 in., 8 in.

17. A wire, 130 ft long, is attached to the top of a 115-foot pole. The wire is straight, and touches the ground 50 ft from the base of the pole. Is the pole vertical? How can you decide?

## Spiral Review

**18.** A certain laser printer can print 8 pages per minute. Write and graph a direct variation equation based on this fact. *(Section 3.2)*

**Find the reciprocal of each number.** *(Section 4.2)*

**19.** $\frac{1}{7}$　　　　**20.** 12　　　　**21.** $\frac{-18}{5}$　　　　**22.** −1

# Section 6.2 — Irrational Numbers

**GOAL**

**Learn how to...**
- find noninteger square roots

**So you can...**
- use decimal approximations for numbers such as $\sqrt{17}$

### Application
A pet owner plans to use part of her yard as an exercise region for her pets. If she wants a square exercise region with an area of 20 m², she can use irrational numbers to find the length of the sides of the square.

### Terms to Know / Example / Illustration

| Terms to Know | Example / Illustration |
|---|---|
| **Rational number (p. 246)** — any number that can be written as a ratio of two integers $\frac{a}{b}$, where $b \neq 0$ (When written as a decimal, a rational number either terminates or repeats.) | $\frac{2}{3}$  $0.\overline{4} = \frac{4}{9}$<br>$3$  $\sqrt{3.61} = 1.9$ |
| **Irrational number (p. 246)** — a number that cannot be written as a ratio of two integers (When written as a decimal, an irrational number never terminates or repeats.) | $-\sqrt{3} \approx -1.73205$<br>$\pi \approx 3.141593$<br>$\sqrt{8} \approx 2.828427$ |
| **Real number (p. 246)** — a number that is rational or irrational and can be located on a number line | $\frac{2}{3}$  $0.\overline{4}$  $3$  $\pi$<br>$-\sqrt{3}$  $\sqrt{8}$  $\sqrt{3.61}$ |
| **Dense (p. 246)** — the property of real numbers which states that "for any two real numbers, there is always another real number between them" | The two numbers 35.137 and 35.138 are very close to each other on a number line, but there are many numbers between them. Since 35.137 = 35.1370 and 35.138 = 35.1380, one of the real numbers between them is 35.1377. |

Study Guide, ALGEBRA 1: EXPLORATIONS AND APPLICATIONS
Copyright © McDougal Littell Inc. All rights reserved.

# Understanding the Main Ideas

## Rational, irrational, and real numbers

All the numbers that can be shown on a number line are real numbers. All real numbers are either rational or irrational. No matter how close two real numbers are to each other, you can always find another real number between them.

### Example 1

Name a real number, if possible, that fits each description.

a. not rational
b. rational and an integer
c. rational but not an integer
d. both rational and irrational

### Solution

a. If a number is not rational, then it is irrational. One example is $\sqrt{2}$.
b. All integers are rational numbers since they can be written as fractions with denominator 1. One example is –5.
c. One example of a rational number that is not an integer is $\frac{2}{3}$.
d. No number can be both rational and irrational.

**Tell whether each number is a rational number, an irrational number, an integer, or a real number. (More than one category may apply to a given number.)**

1. $-\frac{2}{3}$
2. $\sqrt{41}$
3. $-\sqrt{64}$
4. $-\sqrt{200}$
5. –5
6. 0

**Name a real number between each pair of numbers.**

7. 2.705 and 2.706
8. $\sqrt{50}$ and $\sqrt{51}$
9. $-\frac{13}{5}$ and $-\frac{13}{4}$

## Calculator values for irrational numbers

The square root key on a calculator gives you decimal approximations for irrational numbers.

### Example 2

a. Find the value of $\sqrt{731}$ to the nearest hundredth.

b. Evaluate $\dfrac{\sqrt{10}}{\sqrt{5}}$ to the nearest hundredth.

c. If $r$ = the radius and $A$ = the area, use the formula $r = \sqrt{\dfrac{A}{\pi}}$ to find the radius of a circle, to the nearest tenth, whose area is 20 m$^2$.

### Solution

a. $\sqrt{731} \approx 27.04$

b. $\dfrac{\sqrt{10}}{\sqrt{5}} \approx \dfrac{3.1622777}{2.236068} \approx 1.41$

c. $r = \sqrt{\dfrac{A}{\pi}} \approx \sqrt{\dfrac{20}{3.1416}} \approx \sqrt{6.3662} \approx 2.52$ m

**Use a calculator to find each value to the nearest hundredth.**

10. $\sqrt{2000}$

11. $-\sqrt{\dfrac{1}{2}}$

12. $\sqrt{7} + \sqrt{8} + \sqrt{9}$

**Arrange each group of numbers in ascending order from smallest to largest. (*Hint*: Use a calculator to find a decimal approximation for each number.)**

13. $\dfrac{17}{2}$, 8, $\sqrt{65}$, $\dfrac{119}{15}$, $\sqrt{63.5}$

14. $\sqrt{100}$, $\sqrt{109}$, 10.3, $\dfrac{198}{20}$

**Find the radius, to the nearest tenth, of a circle whose area is given. Use the formula $r = \sqrt{\dfrac{A}{\pi}}$, where $r$ = the radius and $A$ = the area of a circle. Use $\pi \approx 3.1416$.**

15. 15.32 in.$^2$

16. 3.1416 ft$^2$

## *Repeating decimals*

The number $2.6\overline{7}$ represents the repeating decimal 2.6777… . Every repeating decimal, such as 0.414141… or $2.6\overline{7}$, is a rational number because a repeating decimal can be written as a fraction.

### Example 3

Write 0.414141… as a fraction in lowest terms.

### Solution

Write 0.414141... as $0.\overline{41}$.

Let $b = 0.\overline{41}$ ← Use a variable to represent the repeating decimal.

$100b = 100(0.\overline{41})$ ← There are 2 digits in the repeating block, so multiply both sides by 100.

$100b = 41.\overline{41}$
$\underline{-b = -0.\overline{41}}$ ← Subtract $b = 0.\overline{41}$ from the equation above. Notice that this step eliminates the decimal part of the number.
$99b = 41$

$b = \dfrac{41}{99}$ ← Divide both sides by 99.

So, $0.\overline{41} = \dfrac{41}{99}$.

---

**For Exercises 17–19, write each repeating decimal as a fraction in simplest terms.**

**17.** 0.777...  **18.** $0.2\overline{4}$  **19.** $0.\overline{008}$

**20. Mathematics Journal** Explain one way to find the rational number that is halfway between two given rational numbers.

............................
### Spiral Review

**For each pair of similar figures, write the ratio of the perimeters. Find the ratio of the areas.** *(Section 5.4)*

**21.**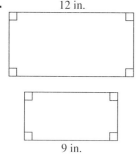

**22.**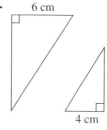

**The legs of a right triangle have lengths $a$ and $b$. The hypotenuse has length $c$. Find the length of the missing side of each right triangle.** *(Section 6.1)*

**23.** $a = 6$ km, $b = 8$ km  **24.** $b = 15$ in., $c = 17$ in.

**25.** $a = 9$ cm, $c = 41$ cm  **26.** $a = 3$ ft, $b = 3$ ft

# Section 6.3 Calculating with Radicals

**GOAL**

**Learn how to . . .**
- multiply and divide radicals
- simplify radicals

**So you can . . .**
- use square roots to solve area and other problems

## Application

A farmer leases the two square fields shown in the diagram below. The fields are on opposite sides of a road, and the side of one field aligns with the opposite side of the other. The farmer can use radical numbers to find the total length of the road that borders the two fields.

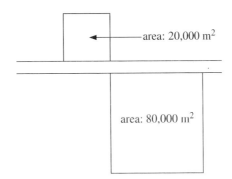

## UNDERSTANDING THE MAIN IDEAS

### Simplifying radicals

An expression containing radicals is in simplest form when (1) none of the numbers under a radical sign has a perfect-square factor, (2) there are no fractions under a radical sign, and (3) there are no radicals in any denominators.

The properties below are used to simplify radicals.

*Properties of Square Roots*

For positive numbers $a$ and $b$,

1. $\sqrt{ab} = \sqrt{a} \cdot \sqrt{b}$
2. $\sqrt{\dfrac{a}{b}} = \dfrac{\sqrt{a}}{\sqrt{b}}$

### Example 1

Write each expression in simplest form.

a. $\sqrt{200}$   b. $\sqrt{\dfrac{100}{64}}$   c. $\sqrt{75} + \sqrt{12}$

---

**Study Guide,** ALGEBRA 1: EXPLORATIONS AND APPLICATIONS
Copyright © McDougal Littell Inc. All rights reserved.

### Solution

**a.** $\sqrt{200} = \sqrt{100 \cdot 2}$ ← 100 is a perfect-square factor of 200.
$= \sqrt{100} \cdot \sqrt{2}$ ← Use property 1.
$= 10\sqrt{2}$

**b.** $\sqrt{\dfrac{100}{64}} = \dfrac{\sqrt{100}}{\sqrt{64}}$ ← Use property 2.
$= \dfrac{10}{8}$
$= \dfrac{5}{4}$

**c.** $\sqrt{75} + \sqrt{12} = \sqrt{25 \cdot 3} + \sqrt{4 \cdot 3}$
$= \sqrt{25} \cdot \sqrt{3} + \sqrt{4} \cdot \sqrt{3}$ ← Use property 1.
$= 5\sqrt{3} + 2\sqrt{3}$
$= (5 + 2)\sqrt{3}$ ← Use the distributive property.
$= 7\sqrt{3}$

**Write each expression in simplest form.**

1. $\sqrt{108}$
2. $\sqrt{\dfrac{121}{100}}$
3. $7\sqrt{5} - \sqrt{45}$
4. $\sqrt{\dfrac{18}{81}}$
5. $4\sqrt{3} \cdot 2\sqrt{27}$
6. $\sqrt{50} + \sqrt{128}$

## *Estimating square roots*

To estimate the square root of an integer that is not a perfect square, we can use the two integer perfect squares closest to it, one on each side.

### Example 2

Estimate $\sqrt{105}$ within a range of two integers.

### ■ Solution ■

Find the perfect square just less than 105 and the first perfect square just greater than 105.

$$9^2 = 81 \qquad 10^2 = 100 \qquad 11^2 = 121$$

So the next smallest perfect square is 100 and the next largest perfect square is 121.

Since $\quad 100 \quad < \quad 105 \quad < \quad 121$

then $\quad \sqrt{100} \quad < \quad \sqrt{105} \quad < \quad \sqrt{121}$

or $\quad 10 \quad < \quad \sqrt{105} \quad < \quad 11$

So, the square root of 105 is between 10 and 11.

---

**For Exercises 7–9, estimate each number within a range of two integers.**

**7.** $\sqrt{10}$      **8.** $\sqrt{50}$      **9.** $\sqrt{175}$

**10.** The area of a square is 230 ft². Between what two whole-number lengths is the length of each side of the square?

**11. Open-ended Problem** Find the dimensions of a rectangular playing field at or near your school. What is the area of the field? What would be the length of the side of a square that has the same area?

·············

## Spiral Review

**For Exercises 12–15, solve each equation.** *(Section 4.3)*

**12.** $5(x + 17) = 35$

**13.** $6 + 3x = 8x - 9$

**14.** $5x = 70 - 9x$

**15.** $3(x - 4) = 5x + 16$

**16.** $\triangle XYZ \sim \triangle RST$. Find each missing side length and angle measure. *(Section 5.3)*

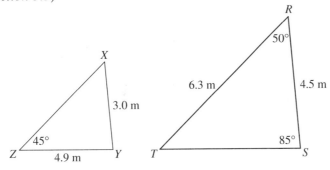

# Section 6.4: Multiplying Monomials and Binomials

**GOAL**

**Learn how to...**
- find products of monomials and binomials

**So you can...**
- find the area of a complex figure

## Application

An artist wants to place a mat around a picture. The dimensions of the picture are 20 in. by 24 in. The mat will extend $x$ inches beyond the picture on both the left and the right, and will extend $y$ inches beyond the picture on both the top and the bottom. The artist can use binomials and their products to find the area of the mat.

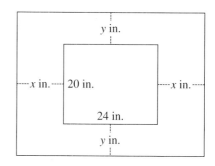

### Terms to Know | Example / Illustration

| Terms to Know | Example / Illustration |
|---|---|
| **Monomial (p. 260)** an expression with one term | $xyz$  $2x^2y^2z$  $5b^2$ |
| **Binomial (p. 260)** an expression with two unlike terms | $x + y$  $5b^2 + 32$  $2x^2y^2 - x^2y^2z^2$ |
| **Trinomial (p. 260)** an expression with three unlike terms | $x - y + 1$  $5b^2 + 2b + 32$  $2x^2 - 2y^2 - 2z^2$ |

## UNDERSTANDING THE MAIN IDEAS

### Terms and expressions

A variable expression can be named by the number of terms in the expression. Each term can be a constant, a variable, or the product of a constant and one or more variables. The terms in a variable expression are separated by the operation symbols + and −. There are four terms in the expression below.

$$5xy + 7pq - 13x + pq$$

Notice that two of the terms have the same variable part. These terms, $7pq$ and $pq$, are called *like terms*, and they can be combined using the distributive property.

$$7pq + pq = (7 + 1)pq = 8pq$$

Study Guide, ALGEBRA 1: EXPLORATIONS AND APPLICATIONS
Copyright © McDougal Littell Inc. All rights reserved.

### Example 1

Simplify each expression. Then tell whether the resulting expression is a *monomial, binomial*, or *trinomial*.

a. $5x + 3y + 8x - 7y$
b. $x + 2x^2 + 3xy$
c. $5x^2z - 7x^2z + 10x^2z$
d. $x - 3y + 4z - 7 + 2x - 5z$

### ■ Solution ■

a. There are two pairs of like terms: $5x$ and $8x$, and $3y$ and $(-7y)$.

$$5x + 3y + 8x - 7y = 5x + 8x + 3y - 7y$$
$$= 13x - 4y$$

The expression $13x - 4y$ has two unlike terms, so it is a binomial.

b. The expression $x + 2x^2 + 3xy$ has no like terms, so it is already simplified. Since it has three terms, it is a trinomial.

c. All three terms are like terms.

$$5x^2z - 7x^2z + 10x^2z = (5 - 7 + 10)x^2z \quad \leftarrow \text{Use the distributive property.}$$
$$= 8x^2z$$

The expression $8x^2z$ has one term, so it is a monomial.

d. $x - 3y + 4z - 7 + 2x - 5z = x + 2x - 3y + 4z - 5z - 7$
$$= 3x - 3y - z - 7$$

This is a four-term expression; it is neither a monomial, a binomial, nor a trinomial.

**Simplify each expression. Then whether each resulting expression is a** *monomial, binomial, trinomial,* **or** *none of these.*

1. $2x^2 + 3xy - 5y^2 + 6x^2 - y^2$
2. $np^2 - 5np^2 + 12np^2$
3. $a + 5a^2 - 3a^3 + 4a + a^2 - a^3$
4. $5 + 5b$

## *Multiplying monomials and binomials*

We often need to find products such as $5(2x + 3)$ and $(x + 3)(x + 2)$. The first expression is the product of a monomial and a binomial. The second expression is the product of two binomials. Products of these types can be modeled with algebra tiles.

### Example 2

Use algebra tiles to find each product.

a. $5(2x + 3)$
b. $(x + 3)(x + 2)$

**Study Guide,** ALGEBRA 1: EXPLORATIONS AND APPLICATIONS
Copyright © McDougal Littell Inc. All rights reserved.

### Solution

**a.** Arrange algebra tiles in a rectangle so that the width is 5 and the length is $(2x + 3)$. By its area, the model shows the product $5(2x + 3)$.

Since there are 10 $x$-tiles and 15 1-tiles, the product $5(2x + 3)$ is $10x + 15$.

**b.** Arrange algebra tiles in a rectangle so that the width is $(x + 3)$ and the length is $(x + 2)$. By its area, the model shows the product $(x + 3)(x + 2)$.

Since there is 1 $x^2$-tile, 5 $x$-tiles, and 6 1-tiles, the product $(x + 3)(x + 2)$ is $x^2 + 5x + 6$.

**Use algebra tiles to find each product. Make a drawing of each tile model and show its dimensions.**

**5.** $3x(x + 5)$

**6.** $(x + 3)(3x + 1)$

Not all products can be modeled conveniently using algebra tiles. These products can be found using the distributive property.

### Example 3

Use the distributive property to find each product.

**a.** $5(2x - 3)$

**b.** $(3x - 2)(4x - 3)$

### Solution

**a.** Use the distributive property to multiply 5 times both terms in the expression $(2x - 3)$.

$$5(2x - 3) = 5 \cdot 2x - 5 \cdot 3$$
$$= 10x - 15$$

**b.** Multiply $3x$ times both terms in the expression $(4x - 3)$. Then multiply $-2$ times the terms in $(4x - 3)$.

$$(3x - 2)(4x - 3) = 3x(4x - 3) + (-2)(4x - 3)$$
$$= 3x(4x) + 3x(-3) + (-2)(4x) + (-2)(-3)$$
$$= 12x^2 - 9x - 8x + 6$$
$$= 12x^2 - 17x + 6$$

**For Exercises 7–10, find each product.**

**7.** $6a(5a - 7)$

**8.** $(2t - 3)(t + 5)$

**9.** $(3 - 2\sqrt{2})(4 + 5\sqrt{2})$

**10.** $(2x + 7)(4x - 5)$

**11.** The length of a rectangle is represented by the expression $(5a + 2)$ and its width is represented by $(3a - 1)$. Find an expression for the area of the rectangle.

**12.** The diagram at the right shows a rectangular walkway around a square pool. Find an expression for the area of the walkway.

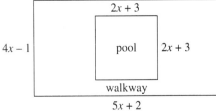

**13. Open-ended Problem** Find binomials for each of the following conditions.

   **a.** two binomials whose product is a trinomial

   **b.** two binomials whose product is a binomial

   **c.** two binomials whose product is a monomial

   **d.** two binomials whose product has four terms

## Spiral Review

**Exercises 14 and 15 refer to a spinner that has ten equal sectors, numbered 1 through 10.** *(Section 5.5)*

**14.** What is the theoretical probability of spinning a number less than 4?

**15.** What is the theoretical probability of spinning an even number that is greater than 7?

**Use an equation to model each set of data.** *(Section 3.6)*

**16.**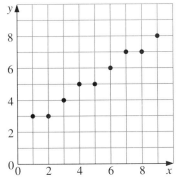

**17.**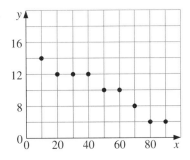

**Study Guide,** ALGEBRA 1: EXPLORATIONS AND APPLICATIONS
Copyright © McDougal Littell Inc. All rights reserved.

# Section 6.5 Finding Special Products

**GOAL**

**Learn how to . . .**
- recognize some special products of binomials

**So you can . . .**
- find products of some variable expressions more quickly

## Application

A square patio is increased in size, with the new patio also being a square. You can use the expression $(x + y)^2$ to find the area of the new patio.

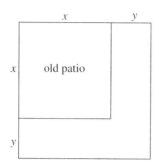

### Terms to Know / Example / Illustration

| Terms to Know | Example / Illustration |
|---|---|
| **Perfect square trinomial** (p. 265) <br> the square of a binomial | $x^2 + 2x + 1 = (x + 1)^2$ <br> $4m^2 + 12m + 9 = (2m + 3)^2$ |
| **Difference of two squares** (p. 266) <br> the product of the sum and difference of two terms | $x^2 - 25 = x^2 - 5^2$ <br> $25a^2 - 64 = (5a)^2 - 8^2$ |

## UNDERSTANDING THE MAIN IDEAS

### Squaring a binomial

Recall that a binomial is an expression with two unlike terms, such as $x + 3$, or $5a - 2$, or $8 - 3\sqrt{2}$. So the expressions $(x + 3)^2$, $(5a - 2)^2$, and $(8 - 3\sqrt{2})^2$ are the square of a binomial.

**Example 1**

Find each product.

a. $(x + 3)^2$  b. $(5a - 2)^2$  c. $(8 - 3\sqrt{2})^2$

### Solution

**a.** Since $(x + 3)^2 = (x + 3)(x + 3)$, use algebra tiles to model the product as a square with side lengths $(x + 3)$. Then find the total area.

There is 1 $x^2$-tile, 6 $x$-tiles, and 9 1-tiles, so $(x + 3)^2 = x^2 + 6x + 9$.

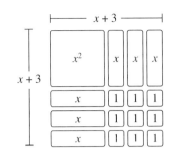

**b.** $(5a - 2)^2 = (5a - 2)(5a - 2)$ &larr; Use the distributive property.
$= 5a(5a - 2) + (-2)(5a - 2)$
$= 25a^2 - 10a - 10a + 4$ &larr; Combine like terms.
$= 25a^2 - 20a + 4$

**c.** $(8 - 3\sqrt{2})^2 = (8 - 3\sqrt{2})(8 - 3\sqrt{2})$ &larr; Use the distributive property.
$= 8(8 - 3\sqrt{2}) + (-3\sqrt{2})(8 - 3\sqrt{2})$
$= 64 - 24\sqrt{2} - 24\sqrt{2} + 9\sqrt{4}$ &larr; $\sqrt{2} \cdot \sqrt{2} = \sqrt{2 \cdot 2}$
$= 64 - 48\sqrt{2} + 18$
$= 82 - 48\sqrt{2}$

**Find each product.**

1. $(4x + 1)^2$
2. $(x + 8)^2$
3. $(11 - 3x)^2$
4. $(2y - 9)^2$
5. $(7 + 3x)^2$
6. $(6 + 3\sqrt{5})^2$
7. $(4 - \sqrt{7})^2$
8. $(4 - 9z)^2$
9. $(3\sqrt{7} - 5)^2$

## The product $(a + b)(a - b)$

When you see an expression that has the form $(a + b)(a - b)$, you can use mental math to write the product. This product has the form $a^2 - b^2$, and is called the *difference of two squares*.

### Example 2

Find the product $(2x + 5)(2x - 5)$.

> **Solution**
>
> To calculate $(2x + 5)(2x - 5)$, use the pattern $(a + b)(a - b) = a^2 - b^2$. In the pattern, replace $a$ with $2x$ and replace $b$ with 5.
>
> $$(a + b)(a - b) = a^2 - b^2$$
> $$(2x + 5)(2x - 5) = (2x)^2 - (5)^2$$
> $$= (2x)(2x) - (5)(5)$$
> $$= 4x^2 - 25$$
>
> The product, $4x^2 - 25$, is the difference of two squares.

**For Exercises 10–15, find each product.**

**10.** $(x + 10)(x - 10)$  **11.** $(3y - 4)(3y + 4)$  **12.** $(7 + \sqrt{10})(7 - \sqrt{10})$

**13.** $(2z - 11)(2z + 11)$  **14.** $(8 - 5w)(8 + 5w)$  **15.** $(12 - 5\sqrt{2})(12 + 5\sqrt{2})$

**16.** A square has sides of length 5 in. A rectangle has length 1 in. greater than the side of the square and width 1 in. less than the side of the square.

  **a.** What is the area of the rectangle?

  **b.** Which has greater area, the square or the rectangle? By how many square inches?

**17.** Answer the questions in Exercise 16 for a square with sides of length $n$ in.

**18. Mathematics Journal** Write a letter describing your school or town to someone who is not familiar with it. In the letter, use diagrams to describe buildings, athletic fields, and other property. When is a diagram useful in a description?

## Spiral Review

**Find an equation of the line through the given points.** *(Section 3.5)*

**19.** (3, 1) and (5, 5)   **20.** (–3, 2) and (1, –4)

**Find an equation of the line with the given slope and through the given point.** *(Section 3.5)*

**21.** slope = –6; (5, 1)   **22.** slope = $-\frac{1}{3}$; (6, –6)

**Find each product.** *(Section 6.4)*

**23.** $6t(7t - 9)$   **24.** $(5s + 7)(4s - 1)$   **25.** $(4y - 3)(3 - 4y)$

# Chapter 6 Review

**CHAPTER CHECK-UP**

Complete the exercises for a review of Chapter 6. If you have difficulty with a particular problem, review the indicated section.

1. Find the square root $-\sqrt{64}$. *(Section 6.1)*

2. The length of the hypotenuse of a right triangle is 130 cm and the length of one leg is 120 cm. What is the length of the other leg? *(Section 6.1)*

3. Can 21 in., 28 in., and 36 in. be the lengths of the three sides of a right triangle? Explain your answer. *(Section 6.1)*

4. **Open-ended Problem** In a classroom, use a corner (point $C$) and any two locations (points $A$ and $B$) on the walls that form the corner. Measure the distance from $A$ to $C$, the distance from $B$ to $C$, and the distance from $A$ to $B$. Is $C$ a right angle? How can you use your measurements to tell? *(Section 6.1)*

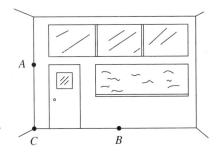

5. Give an example of a real number that is rational but not an integer. *(Section 6.2)*

6. Arrange the numbers 0.54, $0.5\overline{4}$, $0.\overline{54}$, $0.\overline{5}$, 0.45, $0.\overline{4}$, $0.4\overline{5}$, and $0.\overline{45}$ in ascending order from smallest to largest. *(Section 6.2)*

7. Write the repeating decimal 0.727272... as a fraction in lowest terms. *(Section 6.2)*

**For Exercises 8–10, write each expression in simplest form.** *(Section 6.3)*

8. $\sqrt{125} - \sqrt{20}$     9. $9\sqrt{75}$     10. $2\sqrt{6} \cdot \sqrt{98}$

11. Estimate $\sqrt{175}$ within the range of two integers. *(Section 6.3)*

12. Simplify the expression $5xy + 3y^2 - 12xy + y^2 + y$. *(Section 6.4)*

**Find each product.** *(Sections 6.4 and 6.5)*

13. $8x(5 - 3x)$     14. $(6x + 1)(2x + 3)$     15. $(7 + 3\sqrt{2})(4 - 5\sqrt{2})$

16. $(2x - 9)^2$     17. $(3x + 7)(3x - 7)$     18. $(4z + 5)^2$

**SPIRAL REVIEW   Chapters 1–6**

**Simplify each expression.**

1. $\dfrac{4-6}{7(-1+4)}$     2. $|-18| - |11|$     3. $\dfrac{(-7)(3)}{-2}$

**Solve each equation.**

4. $3 - 5x = 22$     5. $3y - 7 = -1$     6. $-5t = 35$

7. $\dfrac{t}{5} = -100$     8. $\dfrac{-3}{5}x = 4$     9. $7 + \dfrac{x}{4} = 15$

**Tell whether or not the ordered pair is a solution of the equation.**

**10.** $y = 3x + 1$; $(2, 7)$  **11.** $y = 1.3x - 5$; $(15, 1)$

**A line contains the two points $A = (1, -2)$ and $B = (3, 10)$.**

**12.** What is the slope of the line?

**13.** What is an equation of the line in slope-intercept form?

**Solve each proportion.**

**14.** $\dfrac{15}{x} = \dfrac{3}{10}$  **15.** $\dfrac{y}{6.4} = \dfrac{4}{3.2}$

**For Exercises 16–18, refer to similar triangles *ABC* and *XYZ* at the right.**

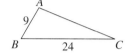

**16.** What is the length of $\overline{AC}$?

**17.** What is the ratio of the perimeter of $\triangle ABC$ to the perimeter of $\triangle XYZ$?

**18.** What is the ratio of the area of $\triangle ABC$ to the area of $\triangle XYZ$?

**19.** When a die is tossed, what is the probability of getting an even number?

**20.** The lengths of the two legs of a right triangle are 24 cm and 32 cm. What is the length of the hypotenuse?

**Simplify each expression.**

**21.** $(3\sqrt{2} - 1)(6\sqrt{2} + 1)$   **22.** $(3 - 5\sqrt{7})^2$   **23.** $(4x + \sqrt{2})(4x - \sqrt{2})$

# Section 7.1 Using Linear Equations in Standard Form

**GOAL**

**Learn how to . . .**
- write and graph equations in standard form

**So you can . . .**
- solve problems with two variables

### Application
An adult's ticket to a museum costs $10 and a child's ticket to the museum costs $5. What combinations of tickets would cost $60?

**Terms to Know** | **Example / Illustration**
---|---
**Standard form of a linear equation (p. 282)** an equation of the form $ax + by = c$, where $a$, $b$, and $c$ are integers, with $a$ and $b$ not both zero | $10x + 5y = 60$
**Horizontal intercept, $x$-intercept (p. 282)** the point of intersection of a graph with the horizontal axis ($x$-axis) |  The $x$-intercept of this graph is 6.
**Vertical intercept, $y$-intercept (p. 282)** the point of intersection of a graph with the vertical axis ($y$-axis) | The $y$-intercept of the graph above is 12.

## UNDERSTANDING THE MAIN IDEAS

A linear equation models a relationship between two variable quantities.

### Example 1

Refer to the situation discussed in the Application. Write an equation that models the possible combinations of adults' and children's tickets that would cost $60.

**Solution**

Let $a$ = the number of adults' tickets and $c$ = the number of children's tickets. Then the equation is $10a + 5c = 60$.

1. Use the equation in Example 1 to find three combinations of tickets that each cost $60.

2. A weekly special at a grocery store offers tomatoes for $1.59 per pound and beans for $1.19 per pound. Write an equation that shows you spent $7.00 on $t$ pounds of tomatoes and $b$ pounds of beans.

3. A music store sells compact discs for $12 and cassette tapes for $8. Write an equation that shows you spent $88 on $d$ compact discs and $t$ cassette tapes.

## *Graphing linear equations given in standard form*

To graph a linear equation that is written in standard form, first find the $x$- and $y$-intercepts of the graph. These are the points where $y = 0$ and $x = 0$, respectively. Then plot the two intercepts and draw the line through them.

### Example 2

Graph the equation $2x + 3y = 12$.

**Solution**

**Step 1** The equation is linear, so find the $x$- and $y$-intercepts.

$x = 0$:  $2(0) + 3y = 12$      $y = 0$:  $2x + 3(0) = 12$
$\phantom{x = 0:\ \ }3y = 12$      $\phantom{y = 0:\ \ 2x + 3(0)}2x = 12$
$\phantom{x = 0:\ \ \ \ }y = 4$      $\phantom{y = 0:\ \ 2x + 3(0)\ \ }x = 6$

The two points $(0, 4)$ and $(6, 0)$ are on the line.

*(Solution continues on next page.)*

## Solution (continued)

**Step 2** Graph the two points and draw a line through them.

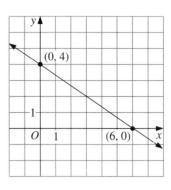

**Graph each equation by using the method in Example 2.**

4. $3x + 4y = 24$
5. $3x - 6y = 12$
6. $2m - 3n = -1$
7. $-4a + 2b = 8$
8. $0.5w - 1.5z = 4.5$
9. $\frac{x}{3} + \frac{y}{2} = 1$

### Example 3

Write the equation $x + 3y = 12$ in slope-intercept form. Then graph the equation.

## Solution

To write the equation in slope-intercept form, solve it for $y$.

$$x + 3y = 12$$
$$x - x + 3y = 12 - x$$
$$3y = 12 - x$$
$$\frac{3y}{3} = \frac{12 - x}{3}$$
$$y = 4 - \frac{1}{3}x, \text{ or } y = -\frac{1}{3}x + 4$$

This slope-intercept form of the equation shows that the $y$-intercept is 4 and the slope is $-\frac{1}{3}$.

*(Solution continues on next page.)*

**Solution** *(continued)*

So, plot the point (0, 4) and then move 1 unit down and 3 units to the right to plot a second point, (3, 3). Now draw the line through the two points. The completed graph is shown at the right.

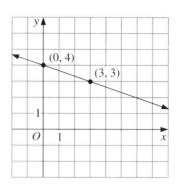

**Rewrite each equation in slope-intercept form. Then graph the equation.**

**10.** $2x + 4y = 12$  **11.** $3x - 5y = 15$  **12.** $3x + 3y = 8$

**13.** $16 = 8x - 2y$  **14.** $0.3x + 0.4y = 1.2$  **15.** $14 - 2y = 7x$

## Spiral Review

**TECHNOLOGY** Graph each equation. Use a graphing calculator if you have one. Use the graph to find the missing value. *(Section 2.6)*

**16.** $y = 2.2x - 1.8$; find the value of $x$ when $y = 5.9$.

**17.** $y = -1.5 - 0.5x$; find the value of $x$ when $y = 0.2$.

**Solve each equation.** *(Section 2.2)*

**18.** $3x - 1 = 12$  **19.** $4x + 1 = \dfrac{1}{4}$  **20.** $15 - 7x = 64$

# Section 7.2 Solving Systems of Equations

**GOAL**

**Learn how to . . .**
- solve systems of equations

**So you can . . .**
- solve problems with two variables

*Application*

The members of the Math Club raised money by holding a car wash. They charged $4 for a car and $6 for a truck. They earned a total of $230 by washing a total of 49 vehicles. You can determine how many cars and how many trucks they washed by writing and solving a system of equations.

### Terms to Know | Example / Illustration

| Terms to Know | Example / Illustration |
|---|---|
| **System of linear equations** (p. 288) — two or more equations relating the same variables | $2x + y = 12$ <br> $3x - 4y = 7$ |
| **Solution of a system** (p. 288) — any pair of numbers that satisfies each equation in a system | The solution of the system above is $(5, 2)$. |

## UNDERSTANDING THE MAIN IDEAS

Systems of equations may be solved graphically, or by the substitution method.

**Example 1**

Solve the system of equations graphically.

$x + 2y = 6$
$2x - y = 2$

---

**Study Guide,** ALGEBRA 1: EXPLORATIONS AND APPLICATIONS
Copyright © McDougal Littell Inc. All rights reserved.

### Solution

**Step 1** Graph the two equations on the same coordinate plane.

The *x*- and *y*-intercepts of the line $x + 2y = 6$ are $(6, 0)$ and $(0, 3)$, respectively. The *x*- and *y*-intercepts of the line $2x - y = 2$ are $(1, 0)$ and $(0, -2)$, respectively.

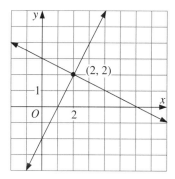

**Step 2** Identify the point where the graphs intersect: $(2, 2)$.

The solution $(x, y)$ is $(2, 2)$.

**Solve each system of equations by graphing.**

1. $3x + 5y = -13$
   $2x - y = 0$

2. $x + 2y = 4$
   $4x - 5y = -23$

3. $2x + 5y = -13$
   $3x - y = 6$

### Example 2

Use substitution to solve the system of equations.

$5x + 3y = 17$
$x - 2y = 6$

### Solution

**Step 1** Solve either equation for one of the variables.

Solve the second equation for *x*: $x - 2y = 6 \rightarrow x = 6 + 2y$

**Step 2** Substitute the result of Step 1 in the other equation.

Substitute $(6 + 2y)$ for *x* in the first equation.

$5x + 3y = 17$

$5(6 + 2y) + 3y = 17$ ← Use the distributive property.

$30 + 10y + 3y = 17$

$30 + 13y = 17$

$13y = -13$

$y = -1$

*(Solution continues on next page.)*

### Solution (continued)

*Step 3* Substitute the result of Step 2 in the equation from Step 1.

Substitute $-1$ for $y$ in $x = 6 + 2y$.

$x = 6 + 2(-1)$
$\phantom{x} = 6 + (-2)$
$\phantom{x} = 4$

*Step 4* Check the results from Steps 2 and 3 in both *original* equations.

$5x + 3y = 17 \qquad\qquad x - 2y = 6$
$5(4) + 3(-1) = 17 \qquad 4 - 2(-1) = 6$
$20 + (-3) = 17 \checkmark \qquad 4 + 2 = 6 \checkmark$

The solution $(x, y)$ is $(4, -1)$.

**Use substitution to solve each system of equations.**

4.  $3x + y = 16$
    $y = x + 4$

5.  $4x + y = 32$
    $x = y + 3$

6.  $3x + y = 10$
    $x + y = 6$

7.  $4x + 3y = 31$
    $2x - y = -7$

8.  $x + 2y = 2$
    $5x - 3y = -29$

9.  $6x - y = 31$
    $4x + 3y = 17$

10. $4x + 5y = 48$
    $3x - y = -2$

11. $2x - 9y = 14$
    $6x - y = 42$

12. $7x - 6y = -30$
    $x - 4y = -20$

### Example 3

Refer to the situation discussed in the Application. How many cars and how many trucks did the Math Club members wash?

### Solution

*Step 1* Write a system of two equations using two variables.

Let $c$ = the number of cars washed and $t$ = the number of trucks washed. Since the total number of cars and trucks washed was 49, one equation is $c + t = 49$. Since they charged $4 for each car and $6 for each truck, a second equation is $4c + 6t = 230$.

*(Solution continues on next page.)*

> ■ **Solution** ■ *(continued)*
>
> **Step 2** Solve the system of equations.
>
> $c + t = 49 \rightarrow t = 49 - c$
>
> $4c + 6t = 230$
>
> $4c + 6(49 - c) = 230 \quad \leftarrow$ Substitute $49 - c$ for $t$.
>
> $4c + 294 - 6c = 230$
>
> $294 - 2c = 230$
>
> $64 = 2c$
>
> $32 = c$
>
> $t = 49 - c \quad \leftarrow$ Substitute 32 for $c$.
>
> $= 49 - 32$
>
> $= 17$
>
> The club members washed 32 cars and 17 trucks.

**13.** Football tickets cost $4 for a reserved seat and $3 for general admission. The total gate receipts for a game at which 217 tickets were sold was $726. How many of each type of ticket were sold?

**14.** Jani has $20 to spend at the store. She can spend all the money for two cassette tapes and a poster, or she can buy one cassette tape and two posters and have $1 left over. What is the price of a cassette tape?

..................
## Spiral Review

**15. Geometry** The Cosmoclock 21 in Yokohama City, Japan, is one of the largest Ferris wheels in the world. Its diameter is 328 ft. The 60 arms holding the gondolas serve as the second hands for the clock.

**a.** How many feet would you travel in one-half revolution of the Ferris wheel? *(Toolbox, p. 593)*

**b.** Find your speed in miles per hour if the wheel makes one revolution every 45 s. Use the fact that 1 mi = 5280 ft. *(Section 3.1)*

# Section 7.3
## Solving Linear Systems by Adding or Subtracting

**GOAL**

**Learn how to...**
- solve systems of equations by adding or subtracting

**So you can...**
- solve problems in two variables

*Application*

Wood Suppliers delivers firewood for a charge per cord of wood plus a delivery fee. If you pay $405 for 1.5 cords of wood and $530 for 2 cords of wood, how much will it cost to have one-half cord delivered?

## UNDERSTANDING THE MAIN IDEAS

Some systems of equations can be solved by adding or subtracting the corresponding sides of the equations if this will eliminate one of the variables. If the coefficients of one of the variables are the same, the system can be solved by subtracting one equation from the other. If the coefficients of one of the variables are opposites, you can solve the system by adding the two equations together.

### Example 1

Identify the method that could be used most easily to solve each system of equations.

**a.** $y = 2x + 4$
$2x + 7y = 3$

**b.** $x + y = 7$
$2x - y = -1$

**c.** $2x + 5y = 7$
$2x - 4y = -2$

### ▬ Solution ▬

**a.** Since one equation is already solved for $y$, use substitution.

**b.** Since the $y$-terms have coefficients that are opposites, adding the equations will eliminate the variable $y$. The resulting equation can then be solved for $x$.

**c.** Since the coefficients of the $x$-terms are the same, subtracting will eliminate the variable $x$. The resulting equation can then be solved for $y$.

**Study Guide,** ALGEBRA 1: EXPLORATIONS AND APPLICATIONS
Copyright © McDougal Littell Inc. All rights reserved.

## Example 2

Solve each system of equations by addition or subtraction.

**a.** $3x + y = 10$
$x - y = -2$

**b.** $4x + 5y = 22$
$2x + 5y = 6$

### ■ Solution ■

**a.** Since the coefficients of the $y$-terms are opposites, add the corresponding sides of the two equations.

$$3x + y = 10$$
$$+ (x - y) = + (-2)$$
$$4x = 8$$

$$\frac{4x}{4} = \frac{8}{4}$$

$$x = 2$$

Now, substitute 2 for $x$ in either equation.

$x - y = -2$
$2 - y = -2$
$-y = -4$
$y = 4$

The solution $(x, y)$ is $(2, 4)$.

**b.** Since the coefficients of the $y$-terms are the same, subtract one equation from the other.

$$4x + 5y = 22$$
$$- (2x + 5y) = - (6)$$
$$2x = 16$$

$$\frac{2x}{2} = \frac{16}{2}$$

$$x = 8$$

Now, substitute 8 for $x$ in either equation.

$2x + 5y = 6$
$2(8) + 5y = 6$
$16 + 5y = 6$
$5y = -10$
$y = -2$

The solution $(x, y)$ is $(8, -2)$.

**Solve each system of equations by adding or subtracting.**

1. $3x - y = 5$
   $2x + y = 15$

2. $4x + y = 10$
   $6x - y = 20$

3. $-30 = 10x + 7y$
   $-24 = 8x + 7y$

4. $5x - y = 22$
   $5x + 4y = -63$

5. $3x - 5y = 61$
   $3x - y = 17$

6. $6x + 11y = -48$
   $x + 11y = -8$

7. $-2x = 7y - 8$
   $4x = -7y + 9$

8. $4x + 9y = 75$
   $4x + 3y = 33$

9. $x - 2y - 15 = 0$
   $-2x + 3y = -11$

**Solve each system by graphing, substituting, adding, or subtracting.**

10. $x = 19 - 4y$
    $6x + 5y = 38$

11. $5x - y = -3$
    $3x + 8y = 24$

12. $x + 2y = -2$
    $5x - 2y = -16$

13. $y = 3x - 6$
    $5y = -10 - 5x$

14. $2x + 3y = 18$
    $x + y = 5$

15. $y = x - 3$
    $4x + y = 32$

> ### Example 3
>
> Refer to the situation discussed in the Application. How much will it cost to have one-half cord of wood delivered?
>
> ### ■ Solution ■
>
> Use the information to write a system of equations.
>
> Let $w$ = the charge for each cord of wood and $d$ = the delivery fee.
> For 1.5 cords of wood, the equation is $1.5w + d = 405$.
> For 2 cords of wood, the equation is $2w + d = 530$.
>
> Subtracting the two equations will eliminate $d$.
>
> $$\begin{aligned} 1.5w + d &= 405 \\ -(2w + d) &= -(530) \\ \hline -0.5w &= -125 \\ -2(-0.5w) &= -2(-125) \\ w &= 250 \end{aligned}$$
>
> Use the value of $w$ to solve for $d$.
>
> $2w + d = 530$
> $2(250) + d = 530$  ← Substitute 250 for $w$.
> $500 + d = 530$
> $d = 30$
>
> The delivery fee is \$30 and the charge per cord of wood is \$250. So for one-half cord you will pay \$125 + \$30, or a total cost of \$155.

**16.** Suppose Wood Suppliers offers a one-week special this fall, one cord of wood for $255 or 2 cords for $485, including delivery. What is the charge for each cord of wood during the special, and what is the new delivery charge?

## Spiral Review

**Solve each equation.** *(Section 4.4)*

**17.** $4.5x + 15.1 = 1.6$

**18.** $\frac{3}{4}x - \frac{1}{4} = 8$

**Graph each equation. Then write an inequality to represent the *x*-values where *y* ≤ 1.** *(Section 4.5)*

**19.** $y = 3x - 2$

**20.** $y = -2x + 1$

# Section 7.4
## Solving Linear Systems Using Multiplication

**GOAL**

**Learn how to . . .**
- write equivalent systems of equations

**So you can . . .**
- solve systems of equations

### Application

For a wall hanging she plans to make, Lynn needs to buy some flower-print fabric and some plain fabric for the trim. She found printed material for $8/yd and plain material for $3/yd. She needs a total of 15 yd of material and she has $100 to spend on it. How much of each material should she buy if she wants to use all of her money? A linear system of equations can be used to answer this question.

### UNDERSTANDING THE MAIN IDEAS

For many systems of equations you cannot eliminate one of the variables by simply adding or subtracting the equations. For these systems, you can use multiplication to rewrite the equations as an *equivalent system* that can then be solved using addition or subtraction.

### Example 1

Solve each system of equations.

a. $5x + 4y = 16$
   $x - 2y = 6$

b. $5x + 3y = 13$
   $2x - 5y = -1$

### Solution

a. If the second equation is multiplied by 2, the resulting equation can be added to the first equation in order to eliminate the variable $y$.

$$2(x - 2y = 6) \rightarrow \quad 2x - 4y = 12$$
$$+ (5x + 4y) = + (16)$$
$$\overline{\phantom{XX}7x \quad = \quad 28}$$
$$x = 4$$

Substitute 4 for $x$ in either original equation.

$x - 2y = 6$
$4 - 2y = 6$
$-2y = 2$
$y = -1$

The solution $(x, y)$ is $(4, -1)$.

*(Solution continues on next page.)*

### Solution (continued)

**b.** For this system, both equations need to be transformed. If the first equation is multiplied by 2 and the second is multiplied by –5, the resulting equations can be added together to eliminate the variable $x$.

$$2(5x + 3y = 13) \rightarrow 10x + 6y = 26$$
$$-5(2x - 5y = -1) \rightarrow +(-10x + 25y) = +(5)$$
$$31y = 31$$
$$y = 1$$

Substitute 1 for $y$ in either original equation.

$$5x + 3y = 13$$
$$5x + 3(1) = 13$$
$$5x + 3 = 13$$
$$5x = 10$$
$$x = 2$$

The solution $(x, y)$ is $(2, 1)$.

**Solve each system of equations.**

1. $3x + 5y = -9$
   $6x + y = 9$

2. $-2x + 3y = -4$
   $4x - 2y = 16$

3. $7x + 8y = 23$
   $3x - 2y = -1$

4. $4x - 3y = 4$
   $7x + 2y = 7$

5. $4x + y = 42$
   $6x - 5y = 50$

6. $6x + 5y = 5.1$
   $4x - 2y = -1.4$

Some systems of equations have no solution. When solving these systems, you get a statement that is *never* true. Other systems of equations have many solutions. When solving these systems, you get a statement that is *always* true.

### Example 2

For each system of equations, give the solution or state whether there are *no solutions* or *many solutions*.

**a.** $2x + 3y = 5$
   $4x + 6y = 6$

**b.** $x + 2y = 5$
   $3x + 6y = 15$

### Solution

**a.** Multiply the first equation by –2 and add the resulting equation to the second equation.

$$-2(2x + 3y = 5) \rightarrow \begin{array}{r} -4x - 6y = -10 \\ + (4x + 6y) = + (6) \\ \hline 0 = -4 \end{array}$$

Since it is *never* true that $0 = -4$, there are *no solutions* to this system.

**b.** Multiply the first equation by 3 and subtract the second equation from the result.

$$3(x + 2y = 5) \rightarrow \begin{array}{r} 3x + 6y = 15 \\ - (3x + 6y) = - (15) \\ \hline 0 = 0 \end{array}$$

Since it is *always* true that $0 = 0$, this system has *many solutions*.

---

**For each system of equations, give the solution or state whether there are *no solutions* or *many solutions*.**

7. $x + 7y = 12$
   $2x + 14y = 16$

8. $x - y = 15$
   $3x - 3y = 45$

9. $5 = 2x - 7y$
   $15 = 6x - 7y$

10. $5 = 4a - 3b$
    $15 = 8a - 6b$

11. $x + 2y = 17$
    $3x + y = -2$

12. $125 = 5w + 10z$
    $25 = w + 2z$

## Choosing a strategy

When choosing a method to use when solving a system of equations, follow the guidelines given below.

- When one of the equations is already solved for one of the variables, use substitution.
- When the coefficients of one variable are opposites, use addition.
- When the coefficients of one variable are the same, use subtraction.
- When no corresponding coefficients are the same, use multiplication and then addition or subtraction.
- When you want a visual understanding of the problem, use graphing.

### Example 3

Solve the problem given in the Application.

> **■ Solution ■**
>
> Let $f$ = the number of yards of flower-print fabric and $p$ = the number of yards of plain fabric.
>
> Lynn wants to buy 15 yd of fabric in all. → $f + p = 15$
> She has $100 to spend on the fabric.      → $8f + 3p = 100$
>
> Since no corresponding coefficients are the same, use the multiplication method.
>
> Multiply the first equation by 3 and then subtract the second equation.
>
> $$\begin{array}{rl} 3(f+p=15) \rightarrow & 3f + 3p = 45 \\ & -(8f+3p) = -(100) \\ \hline & -5f = -55 \\ & f = 11 \end{array}$$
>
> Substitute 11 for $f$ in the equation $f + p = 15$.
>
> $11 + p = 15$
> $p = 4$
>
> Lynn should buy 11 yd of flower-print fabric and 4 yd of plain fabric.

**13.** Suppose that Lynn's friend Zenaida has asked Lynn to help her make a wall hanging the same size as Lynn's and using the same two types of fabric, but she only wants to spend $65. How many yards of each material should Zenaida buy?

### Spiral Review

**Graph each equation and each point. Tell whether the ordered pair is a solution of the equation.** *(Section 2.5)*

**14.** $y = 5 - 2x$; (2, 1)

**15.** $y = \frac{1}{3}x + 2$; (3, 5)

**Graph each equation.** *(Section 3.4)*

**16.** $x = -2$

**17.** $y = -x - 4$

**18.** $y = 3$

# Section 7.5 Linear Inequalities

**GOAL**

**Learn how to . . .**
- graph linear inequalities

**So you can . . .**
- visualize problems involving inequalities

### Application

One cup of 1% milk has 2.5 grams of fat. One cup of a brand of frozen yogurt contains 5 grams of fat. If you eat food containing 2000 calories a day, it is recommended that you consume no more than 65 grams of fat. If the 1% milk and frozen yogurt were the only foods you eat each day that contain fat, how many cups of each could you eat without exceeding the 65 grams of fat?

| Terms to Know | Example / Illustration |
|---|---|
| **Linear inequality (p. 307)** an inequality involving two variables | $y < 3x - 5$ |

## UNDERSTANDING THE MAIN IDEAS

To graph a linear inequality, follow these steps.

1. Graph the equation as a solid line if the inequality sign is $\leq$ or $\geq$, and as a dashed line if the inequality sign is $<$ or $>$. This line divides the coordinate plane into two regions.

2. Choose a point on one side of the line and test it in the inequality in order to identify which region of the coordinate plane represents the solutions to the inequality.

3. Shade the appropriate region.

> **Example 1**
>
> Graph each inequality.
>
> **a.** $y \geq \frac{2}{3}x - 1$  **b.** $2x - 5y < 10$

### Solution

**a. Step 1** Graph the equation $y = \frac{2}{3}x - 1$ as a solid line, since points on the line are part of the solution.

**Step 2** Take a test point that is not on the line and substitute it in the inequality. (*Note*: The point (0, 0) is a good choice if it is not on the line.)

$$y \geq \frac{2}{3}x - 1$$

$$0 \geq \frac{2}{3}(0) - 1$$

$$0 \geq -1 \quad \text{true}$$

**Step 3** If the point tested in Step 2 made the inequality true, shade the side of the line where the point is located; otherwise, shade the other side of the line.

The region containing the point (0, 0) should be shaded.

The completed graph is shown at the right.

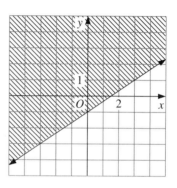

**b. Step 1** To graph the equation $2x - 5y = 10$, first find the $x$- and $y$-intercepts of the graph.

$x$-intercept: if $y = 0$, then $2x = 10$ and $x = 5$; (5, 0)
$y$-intercept: if $x = 0$, then $-5y = 10$ and $y = -2$; (0, -2)

Now, graph the intercepts and draw a dashed line through them, since points on the line are *not* solutions of the inequality.

**Step 2** Test the point (0, 0) in the inequality.

$$2x - 5y < 10$$

$$2(0) - 5(0) < 10$$

$$0 < 10 \quad \text{true}$$

**Step 3** Since (0, 0) is a solution of the inequality, shade the region containing the point (0, 0).

The completed graph is shown at the right.

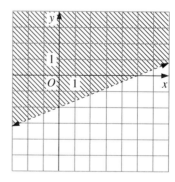

**Match each inequality with its graph.**

1. $y < x - 1$
2. $y > x - 1$
3. $y \leq 1$
4. $x \leq 1$

A.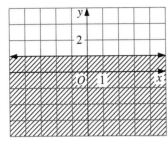

B.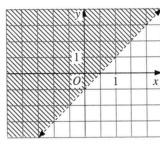

C.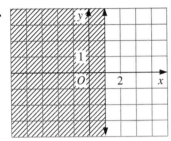

D.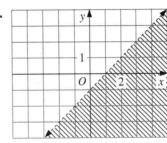

**Graph each inequality.**

5. $y < 3$
6. $x \leq 2$
7. $y \leq 4x$
8. $y > 2x$
9. $y > 3 - x$
10. $y \leq 2x + 3$
11. $y < 4 - x$
12. $y \geq 3x + 2$
13. $3x - 2y \leq 6$
14. $3x + y > 6$
15. $4x - 6y < 12$
16. $-x + 3y < 3$

### Example 2

Write and graph an inequality that can be used to answer the question posed in the Application.

### Solution

Let $x$ = the number of cups of 1% milk and $y$ = the number of cups of frozen yogurt. Then the total grams of fat can be modeled by the inequality $2.5x + 5y \leq 65$ ($x \geq 0$, $y \geq 0$).

The graph of this inequality is shown at the right.

The possible answers to the question are represented by the points in the shaded region of the graph. Three of these combinations are: no milk and 13 cups yogurt, 4 cups milk and 8 cups yogurt, and 8 cups milk and 4 cups yogurt.

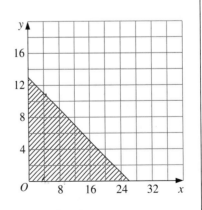

**Study Guide, ALGEBRA 1: EXPLORATIONS AND APPLICATIONS**

**17.** One cup of 2% milk contains 5 grams of fat. How many cups of 2% milk and frozen yogurt containing 5 grams of fat per cup will give you at most 65 grams of fat each day?

## Spiral Review

**Find each product.** *(Section 6.4)*

**18.** $2x(3 - 2x)$　　　**19.** $(2x + 5)(x - 2)$　　　**20.** $(1 + \sqrt{5})(1 - \sqrt{5})$

**Use substitution to solve each system of equations.** *(Section 7.2)*

**21.** $y = x$
　　$2x + 5y = 21$

**22.** $3a - b = -1$
　　$2a + 5b = 5$

**23.** $3x - 2y = 2$
　　$-x + 3y = -17$

# Section 7.6

# Systems of Inequalities

**GOAL**

**Learn how to ...**
- graph systems of inequalities

**So you can ...**
- make decisions when there are many restrictions

## Application

Dry dog food costs $3.00 for a 50 oz bag. Canned dog food costs $.80 for an 8 oz can. Reeta's dog eats at least 100 oz of food a week. She has a weekly budget, and wants to keep the total cost of dog food below $9.00. How much of each type of dog food can Reeta buy and stay within her budget?

### Terms to Know

| | Example / Illustration |
|---|---|
| **System of inequalities (p. 312)**<br>two or more inequalities relating the same variables | $d + 10c \geq 20$<br>$4d + 3c \leq 10$ |
| **Solution of a system of inequalities (p. 312)**<br>all ordered pairs of numbers that satisfy each inequality in a system | 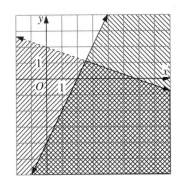<br>The points in the double-shaded region are solutions of the system of inequalities $x + 3y < 6$ and $2x - y \geq 4$. |

### UNDERSTANDING THE MAIN IDEAS

To solve a system of inequalities, graph each inequality on the same coordinate plane. The solution of the system is all the points in the region where the graphs overlap.

**Study Guide,** ALGEBRA 1: EXPLORATIONS AND APPLICATIONS
Copyright © McDougal Littell Inc. All rights reserved.

### Example 1

Graph the system of inequalities.

$$y < 2$$
$$x + y \leq 5$$

**Solution**

*Step 1* Graph the line $y = 2$ as a dashed line and shade the side where $y < 2$.

*Step 2* On the same coordinate plane, graph the line $x + y = 5$ as a solid line and shade the side where $x + y < 5$.

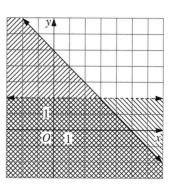

*Step 3* The solution region is the double-shaded part of the figure. Check a point from this region in each inequality.

Using the point (0, 0):  $0 < 2$ ✓
$0 + 0 \leq 5$ ✓

Since the inequalities are true, the correct region is shown.

**Graph each system of inequalities.**

1. $y < 1$
   $x \leq 3$

2. $x > -1$
   $y < x + 1$

3. $y \leq 3 - x$
   $y \geq -2$

4. $x + y > -1$
   $x + y < 4$

5. $x + 2y < 5$
   $x - 2y > 5$

6. $y < 2 + x$
   $y > 1 - 2x$

7. $x + y \geq -2$
   $y \leq 2x$

8. $2x + 3y < 6$
   $y < x + 1$

9. $3x + y > -3$
   $3x - y \leq 3$

### Example 2

Write and graph a system of inequalities to answer the question posed in the Application.

## Solution

Let $x$ = the number of bags of dry dog food and $y$ = the number of cans of moist dog food.

Reeta wants the cost to be below $9.00. → $3x + 0.80y < 9$
Her dog eats at least 100 oz of food each week. → $50x + 8y \geq 100$
The number of bags and cans cannot be negative. → $x \geq 0, y \geq 0$

The graph of the system of inequalities is shown at the right.

Since only whole numbers of cans and bags can be purchased, locate the points in the solution region whose coordinates are both whole numbers:

(2, 3), (2, 2), (2, 1), (2, 0), (1, 7)

Reeta could buy 2 bags of dry food and 3 cans of moist food, or 2 bags and 2 cans, or 2 bags and 1 can, or 2 bags and 0 cans, or 1 bag and 7 cans.

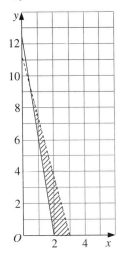

Suppose Reeta discovers that her dog likes a more expensive brand of dry dog food that costs $3.50 for a 50 oz bag. In order to feed her dog well, she increases her budget for dog food to $10 each week.

**10.** Write a system of inequalities for this new situation.

**11.** Graph the system of inequalities from Exercise 10.

**12.** List three possible choices that Reeta can make when she buys dog food.

........................
### Spiral Review

**Solve each proportion.** *(Section 5.1)*

**13.** $\dfrac{9}{12} = \dfrac{x}{2}$     **14.** $\dfrac{m}{7} = \dfrac{12}{8}$     **15.** $\dfrac{3n+1}{4} = \dfrac{5}{2}$

**Find each product.** *(Section 6.5)*

**16.** $(2x - 3)^2$     **17.** $(3 - \sqrt{5})^2$     **18.** $(\sqrt{3} + x)(\sqrt{3} - x)$

# Chapter 7 Review

**CHAPTER CHECK-UP**

Complete these exercises for a review of Chapter 7. If you have difficulty with a particular problem, review the indicated section.

**Graph each equation.** *(Section 7.1)*

1. $2x - 6y = 18$
2. $7x - 4y = 1$
3. $3x - 2y = 12$
4. $2x + 5y = 14$

Kento bought tickets to a concert for a group of people. The tickets cost $8.50 for adults and $5.00 for students. He bought 8 tickets at a total cost of $61. Let *a* = the number of adult tickets and *s* = the number of student tickets. *(Section 7.2)*

5. Write an equation showing that Kento bought a total of 8 tickets.
6. Write an equation showing that Kento spent a total of $61.
7. Solve the system of equations from Exercises 5 and 6 to show how many of each type of ticket Kento bought.

**For each system of equations in Exercises 8–16, give the solution or state whether there are *no solutions* or *many solutions*.** *(Sections 7.2, 7.3, and 7.4)*

8. $2x + y = 5$
   $x - y = 7$
9. $y = 2x + 8$
   $3x + y = -7$
10. $y = 5x - 7$
    $2x - y = -2$
11. $3x - 2y = 11$
    $5x - y = 2$
12. $x + 2y = 8$
    $3x - y = 10$
13. $2x - y = 5$
    $-6x + 3y = -15$
14. $3x - 5y = -2$
    $5x - 3y = 2$
15. $4x - 5y = 34$
    $3x + 6y = 6$
16. $-2x - 7y = -62$
    $10x + 5y = -50$

17. The SuperMarket sells spring water for $1.25 per gallon and seltzer water for $2.25 per gallon. During one day, they sold a total of 89 gal of bottled water, collecting $146.25 in receipts. How many gallons of each type of water did they sell? *(Section 7.4)*

**Graph each inequality.** *(Section 7.5)*

18. $y < 2x - 5$
19. $y \geq \frac{3}{5}x + 3$
20. $4x - 3y > 12$

**For Exercises 21–23, graph each system of inequalities.** *(Section 7.6)*

21. $2x + y < -4$
    $4x + 5y \geq -20$
22. $y > x$
    $2x + 3y > 4$
23. $y > \frac{3}{4}x - 1$
    $y < 2x + 2$

24. Dry cat food costs $2.00 for a 40 oz bag. Moist cat food costs $.75 for a 6 oz can. Chato's cat eats at least 70 oz of food a week. He wants to keep the total weekly cost of the cat food below $6.00. How much of each type of food should be buy? *(Section 7.6)*

**SPIRAL REVIEW    Units 1-7**

Solve each inequality. Graph the solution on a number line.

**1.** $3x - 1 < 8$      **2.** $2(x - 2) \geq 3x$      **3.** $4 - 3x < 2(x + 1)$

Find an equation of the line through the given points.

**4.** $(-3, -2)$ and $(1, 6)$      **5.** $(1, 1)$ and $(1, -6)$

**6.** $(0, 0)$ and $\left(\frac{1}{3}, \frac{1}{4}\right)$      **7.** $\left(-\frac{1}{2}, -2\right)$ and $\left(\frac{3}{2}, -6\right)$

Solve each equation. Round answers to the nearest hundredth.

**8.** $2.67x - 3.25 = 5.65$      **9.** $4.51m + 12.86 = 2.75m$

Evaluate each expression for $a = -2$, $b = 10$, and $c = 0.8$.

**10.** $a^2 + b^2$      **11.** $b^2 - a^2$      **12.** $c^2 - 2ab$

Graph each equation.

**13.** $3x - 4y = 8$      **14.** $15 - 3x = 5y$      **15.** $4x + 3y = 12$

Tell whether the data show direct variation. If so, give the constant of variation and write an equation.

**16.**

| Time (s) | Distance (ft) |
|---|---|
| 5 | 12.5 |
| 10 | 25 |
| 15 | 37.5 |
| 20 | 50 |

**17.**

| Time (s) | Height (ft) |
|---|---|
| 2 | 236 |
| 3 | 306 |
| 5 | 350 |
| 7 | 266 |

# Section 8.1 Nonlinear Relationships

**GOAL**

**Learn how to...**
- analyze the shape of a graph

**So you can...**
- decide whether a relationship is linear

## Application

The Botany Club is planning a plant sale. They estimate that if geraniums are priced at $3.00 they will sell about 400 plants. Based on last year's sale, for each increase of $.25 in the price, 20 fewer plants will be sold. What price should they charge in order to make the most money?

### Terms to Know

**Nonlinear equation (p. 326)**
an equation that models a relationship whose graph is not a line

### Example / Illustration

The graph of the equation $A = \pi r^2$ is not a line.

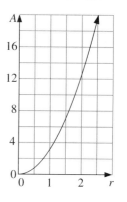

## UNDERSTANDING THE MAIN IDEAS

If the graph of an equation is a line, then the relationship is linear; if it is not a line, the relationship is nonlinear.

### Example 1

Graph each equation. Tell whether each equation is linear or nonlinear.

a. $y = -3x + 2$     b. $y = 3x^2$     c. $y = |x| + 1$

170     Study Guide, ALGEBRA 1: EXPLORATIONS AND APPLICATIONS
Copyright © McDougal Littell Inc. All rights reserved.

### Solution

a. 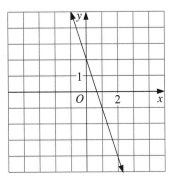   The equation is linear.

b. 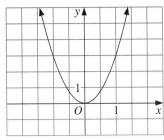   The equation is nonlinear.

c. 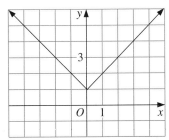   The equation is nonlinear.

**Graph each equation. Tell whether each equation is *linear* or *nonlinear*.**

1. $y = 2x$
2. $y = 3|x|$
3. $y = 2x^2$
4. $y = -3|x|$
5. $y = -2x - 1$
6. $y = |x| - 2$
7. $y = \frac{2}{3}x + 1$
8. $y = 2x^2 - 1$
9. $y = x^2 + 1$

### Example 2

Refer to the Application. What price should the Botany Club charge for each plant? What is their expected total income?

### ■ Solution ■

Complete a table that shows the price per plant, the number of plants sold, and the total income.

| Price per plant | Number of plants sold | Total income |
|---|---|---|
| $3.00 | 400 | $1200 |
| $3.25 | 380 | $1235 |
| $3.50 | 360 | $1260 |
| $3.75 | 340 | $1275 |
| $4.00 | 320 | $1280 |
| $4.25 | 300 | $1275 |
| $4.50 | 280 | $1260 |
| $4.75 | 260 | $1235 |
| $5.00 | 240 | $1200 |

Notice that the total income increases as the price increases to $4.00 and then decreases for higher prices. Since the club wants to make as much money as possible, they should charge $4.00 for each plant, which is expected to result in a total income of $1280.

---

**Suppose a member of the Botany Club found a mistake in their information about last year's sale. The correct information shows that a price of $3.00 gave them sales of 300 geraniums, not 400. It is correct that for each increase of $.25 in the price, 20 fewer plants were sold.**

**10.** Make a table for this situation.

**11.** Make a scatter plot with *price per plant* on the horizontal axis and *total income* on the vertical axis.

**12.** Do you think the relationship between price and income is *linear* or *nonlinear*? Explain your answer.

**13.** What price would you recommend they charge? Explain your reasoning.

· · · · · · · · · · · · · · · · · · · · ·
### Spiral Review

**Find each product.** *(Section 6.4)*

**14.** $2x(12x - 13)$

**15.** $(y + 6)(y - 2)$

**16.** $(3x + 5)(5x - 1)$

**17.** $(3 + \sqrt{3})(2 - \sqrt{3})$

**Rewrite each equation in slope-intercept form. Then graph it. Use a graphing calculator if you have one.** *(Section 7.1)*

**18.** $2x - 7y = 9$

**19.** $2x + 7y = -9$

**20.** $7y - 2x = 9$

# Section 8.2 Exploring Parabolas

**GOAL**

**Learn how to . . .**
- recognize characteristics of parabolas

**So you can . . .**
- predict the shape of a parabola

## Application

A radio telescope has an antenna in the shape of a parabolic dish. The equation that approximates the shape of the dish has the form $y = ax^2$.

### Terms to Know | Example / Illustration

| Terms to Know | Example / Illustration |
| --- | --- |
| **Parabola (p. 331)** a U-shaped curve that is the graph of an equation of the form $y = ax^2$ |  |
| **Quadratic (p. 332)** a function or equation whose graph is a parabola (All quadratic equations have a squared term in them.) | The equation $y = 0.5x^2$ is quadratic. Its graph is shown above. |
| **Vertex of a parabola (p. 332)** the point on a parabola that represents the maximum or minimum value of the function (The plural of vertex is *vertices*.) | The vertex of the parabola shown above is the point $(0, 0)$. The function reaches a *minimum* value there. |
| **Symmetric (p. 333)** the property of a graph exhibited when the graph can be folded so that one half fits exactly over the other half (As a rule, if you draw a vertical line through the vertex of a parabola, the parabola will be symmetric about that line.) | The parabola shown above can be folded along the $y$-axis and the two sides of the parabola will match. |
| **Line of symmetry (p. 333)** a line that divides a graph so that it is symmetric | The $y$-axis is the line of symmetry for the graph of the parabola shown above. |

Study Guide, ALGEBRA 1: EXPLORATIONS AND APPLICATIONS

# UNDERSTANDING THE MAIN IDEAS

A quadratic equation has the form $y = ax^2$ ($a \neq 0$). The graph of a quadratic equation is a parabola that opens upward if $a > 0$ and opens downward if $a < 0$. As the value of $|a|$ increases, the graph of the quadratic function gets narrower.

### Example 1

Predict how the graph of $y = \frac{1}{5}x^2$ will compare with the graph of $y = x^2$. Then sketch the graph of the equation.

**Solution**

The value of $a$ is positive, so the graph will open up just like the graph of $y = x^2$. Also like the graph of $y = x^2$, the vertex of the graph of $y = \frac{1}{5}x^2$ will be $(0, 0)$ and the $y$-axis will be the line of symmetry of the graph. Since $\frac{1}{5} < 1$, the graph will be wider than the graph of $y = x^2$. The graph will go through the point $(5, 5)$ instead of $(5, 25)$.

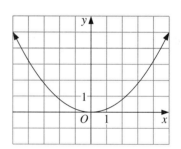

**Match each equation with its graph.**

1. $y = x^2$
2. $y = -\frac{1}{3}x^2$
3. $y = \frac{1}{3}x^2$
4. $y = 3x^2$

A.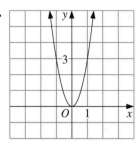

B.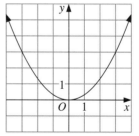

C.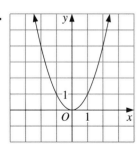

D.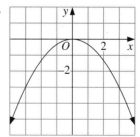

For Exercises 5–10:
   a. Predict how the graph of each equation will compare with the graph of $y = x^2$. Explain how you made your prediction.
   b. Sketch the graph of each equation.

5. $y = 5x^2$
6. $y = -5x^2$
7. $y = -0.25x^2$
8. $y = -\dfrac{1}{2}x^2$
9. $y = \dfrac{1}{4}x^2$
10. $y = 2.5x^2$

### Example 2

A radio telescope has an antenna in the shape of a parabolic dish. The equation that aproximates the shape of the dish has the form $y = ax^2$. Which equation, $y = 600x^2$ or $y = 0.006x^2$, would be more likely to represent the shape of the parabolic dish of a radio telescope?

### ■ Solution ■

The coefficient of $x^2$ would be very small in order to make the parabola as wide as a telescope dish, so the appropriate equation would be $y = 0.006x^2$. (The parabola with equation $y = 600x^2$ would be very narrow.)

11. **Open-ended Problem** Write two equations that might represent the parabolic shape of a radio telescope dish.

12. **Open-ended Problem** Write two equations that would definitely *not* represent the parabolic shape of a radio telescope dish.

13. **Mathematics Journal** In later mathematics courses, parabolas are discussed again under the topic of *conic sections*. Find out what the *focus* of a parabola is and how it is relevant to the use of parabolic dish antennas for radio telescopes.

## Spiral Review

**Find each product.** *(Section 6.5)*

14. $(x - 2)^2$
15. $(y + 7)^2$
16. $(3z + 2)^2$
17. $(a + \sqrt{3})^2$
18. $(b + 4)(b - 4)$
19. $(5x + 3)(5x - 3)$

**Find the unknown length in each right triangle.** *(Section 6.1)*

20.
21.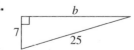
22. Triangle with hypotenuse 35 and leg 12, unknown leg $c$.

# Section 8.3 Solving Quadratic Equations

**GOAL**

**Learn how to...**
- use square roots and graphs to solve simple quadratic equations

**So you can...**
- relate $x$-intercepts and solutions of quadratic equations
- solve simple motion problems

## Application

The distance that a pencil falls after being dropped from the top of a water tower 60 ft high is modeled by the equation $d = 16t^2$, where $d$ = the distance in feet and $t$ = the time in seconds.

### UNDERSTANDING THE MAIN IDEAS

To solve the equation in the form $ax^2 + c = 0$ algebraically, you first solve the equation for $x^2$ and then you take the square root of both sides of the resulting equation to find the value of $x$. (*Note*: When you take the square root of both sides of an equation, you use both the positive and negative values of the square root.) To solve the equation graphically, first graph the function $y = ax^2 + c$ and then find the $x$-intercepts (the points where $y = 0$) of the graph.

### Example 1

Solve the equation $32 = \frac{1}{2}x^2$ algebraically.

**Solution**

$32 = \frac{1}{2}x^2$
$64 = x^2$ ← Multiply both sides by 2.
$\pm\sqrt{64} = x$ ← Find the square root of both sides.
$\pm 8 = x$

The solutions are −8 and 8.

**Solve each equation algebraically.**

1. $x^2 = 64$
2. $144 = m^2$
3. $2x^2 = 32$
4. $-5n^2 = -125$
5. $\frac{1}{3}r^2 = 27$
6. $\frac{2}{3}x^2 = 24$
7. $y^2 - 121 = 0$
8. $0 = a^2 - 16$
9. $p^2 - 0.49 = 0$

### Example 2

Solve the equation $3x^2 + 2 = 50$ by using a graph.

### Solution

**Step 1** Rewrite the equation in the form $ax^2 + c = 0$: $3x^2 - 48 = 0$.

**Step 2** Graph the related equation $y = 3x^2 - 48$.

**Step 3** Since we want the values of $x$ for which $3x^2 - 48 = 0$, identify the points on the graph where $y = 0$. These points are the $x$-intercepts of the graph.

The solutions are $-4$ and $4$.

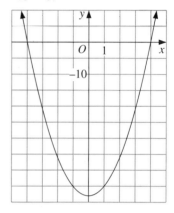

**Solve each equation algebraically or by using a graph.**

10. $x^2 + 2 = 27$
11. $-p^2 + 12 = -52$
12. $3y^2 = 15$
13. $20 = 4n^2$
14. $0 = \frac{1}{9}a^2 - 1$
15. $2x^2 - 20 = 20$
16. $22 = 10m^2 - 8$
17. $-5x^2 + 15 = 5$
18. $\frac{2}{5}a^2 - 7 = 5$

As you have seen, quadratic equations can have two solutions. In real-world situations, you must decide which solutions apply to the given situation and ignore any solution that does not make sense (such as a negative time value).

### Example 3

How long does it take the pencil discussed in the Application to fall 40 ft?

> **Solution**
>
> Graph the equation $y = 16x^2$.
>
> Since $y$ represents the distance, find the point on the graph where $y$ is approximately 40 and $x$ is positive.
>
> Now read the $x$-value for that point: $x \approx 1.6$. This value of $x$ is the time associated with a drop of 40 ft.
>
> It takes about 1.6 s for the pencil to fall 40 ft.
>
>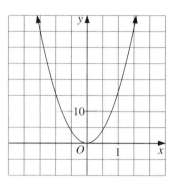

**19.** Use the graph of $y = 16x^2$ to approximate the distance that the pencil will fall in 1.2 s.

**20.** How long will it take for the pencil to hit the ground? (*Hint*: When the pencil hits the ground, it will have traveled 60 ft.)

## Spiral Review

**A house is 70 ft long and 40 ft wide. Find each value.** *(Section 5.2)*

**21.** A floor plan of the house is 8 in. wide. How long is the floor plan?

**22.** A cardboard model of the house is 7 in. wide. How long is the model?

**Find the missing values for each pair of similar figures. *P* represents perimeter and *A* represents area.** *(Section 5.4)*

**23.**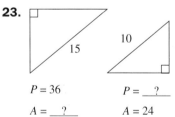
$P = 36$    $P = \underline{\ ?\ }$
$A = \underline{\ ?\ }$    $A = 24$

**24.** 
25 [rectangle]   $P = \underline{\ ?\ }$   $A = 2500$
15 [rectangle]   $P = 150$   $A = \underline{\ ?\ }$

# Section 8.4 Applying Quadratic Equations

## GOAL

**Learn how to . . .**
- use graphs to solve more complicated quadratic equations

**So you can . . .**
- recognize quadratic equations with no solution
- solve any quadratic equation and find heights

### Application

Emile throws a tennis ball from the top of a cliff that is 48 ft high. There is a quadratic relationship between the ball's height above ground level and the time since it was thrown.

### Terms to Know / Example / Illustration

| Terms to Know | Example / Illustration |
|---|---|
| **Quadratic equation (p. 345)** an equation that can be written in the form $ax^2 + bx + c = 0$, where $x \neq 0$ | $2x^2 + 3x - 1 = 0$  $x^2 - 5 = 0$ |

## UNDERSTANDING THE MAIN IDEAS

You can find the solutions of a quadratic equation by using a graph. Another word for the solution of an equation is *root*.

### Example 1

Estimate the solutions of the equation $x^2 - 5x + 3 = 0$ using the graph at the right.

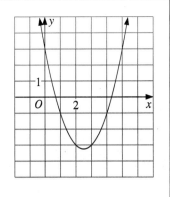

Study Guide, ALGEBRA 1: EXPLORATIONS AND APPLICATIONS
Copyright © McDougal Littell Inc. All rights reserved.

> **Solution**
>
> The $x$-intercepts (the points where $y = 0$) of the graph of $y = x^2 - 5x + 3$ are the solutions of the equation $x^2 - 5x + 3 = 0$.
>
> From the graph, the $x$-intercepts are approximately 0.7 and 4.3.
>
> The solutions are about 0.7 and about 4.3.

**Estimate the solutions of each equation using the given graph. If an equation has no solution, write *no solution*.**

**1.** $-x^2 - 4x - 3 = 0$

**2.** $x^2 + 4x + 4 = 0$

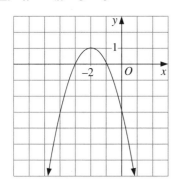
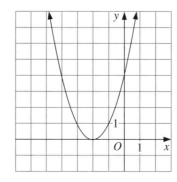

**3.** $-x^2 + 3x - 4 = 0$

**4.** $x^2 + 5x + 3 = 0$

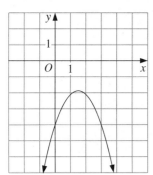
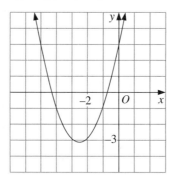

**Use the related graph to solve each equation. If an equation has no solution, write *no solution*.**

**5.** $x^2 + 2x - 3 = 0$
**6.** $2x^2 - 9x + 4 = 0$
**7.** $x^2 - 12x + 36 = 0$
**8.** $x^2 - 2x + 4 = 0$
**9.** $x^2 + 3x - 2 = 0$
**10.** $-3x^2 - 9x + 2 = 0$

> **Example 2**
>
> A tennis ball is thrown from the top of a 48 ft cliff. The quadratic relationship between its height $h$ above ground level and the time $t$ since it was thrown is given by the equation $h = -16t^2 + 16t + 48$. Find the height of the tennis ball after 1.5 s.

### ■ Solution ■

To solve $h = -16t^2 + 16t + 48$ for $t = 1.5$ graphically, begin by graphing the quadratic function $y = -16x^2 + 16x + 48$. (*Note*: Remember that $x$ = the time in seconds and $y$ = the height in feet.)

Then find the $y$-value of the point on the curve where $x = 1.5$.

After 1.5 s, the tennis ball is about 36 ft above ground level.

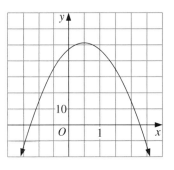

**11.** How many values of $t$ are solutions of the equation $-16t^2 + 16t + 48 = 0$?

**12.** How many of the solutions make sense in the situation described in Example 2? Explain.

**13.** When does the tennis ball first reach the ground?

## Spiral Review

**Write each repeating decimal as a fraction in lowest terms.** (*Section 6.2*)

**14.** 0.5555...      **15.** 0.6363...      **16.** $0.\overline{009}$

**Write each expression in simplest form.** (*Section 6.3*)

**17.** $\sqrt{72}$      **18.** $\sqrt{\dfrac{3}{16}}$      **19.** $\sqrt{\dfrac{81}{25}}$

**20.** $\sqrt{98}$      **21.** $\sqrt{3} + \sqrt{12}$      **22.** $\sqrt{20} + 6\sqrt{5}$

# Section 8.5 The Quadratic Formula

**GOAL**

**Learn how to ...**
- use the quadratic formula

**So you can ...**
- solve quadratic equations without using a graph
- find more precise solutions to real-world problems

## Application

A firefighter aims a hose at a window that is 44 ft above the ground. The equation $h = -0.05d^2 + 2.8d + 5$ describes the path of the water, where $h$ = the height and $d$ = the distance from the firefighter.

### Terms to Know

**Quadratic formula (p. 350)**
the formula used to find the two exact solutions of a quadratic equation in the form $ax^2 + bx + c = 0$

(The quadratic formula is usually written in the form $x = \dfrac{-b \pm \sqrt{b^2 - 4ac}}{2a}$.)

### Example / Illustration

If $3x^2 + 10x + 8 = 0$, then $a = 3$, $b = 10$, and $c = 8$. So,

$$x = \frac{-b \pm \sqrt{b^2 - 4ac}}{2a}$$

$$= \frac{-10 \pm \sqrt{10^2 - 4(3)(8)}}{2(3)}$$

$$= \frac{-10 \pm \sqrt{100 - 96}}{6}$$

$$= \frac{-10 \pm 2}{6}$$

$$x = -2 \text{ or } x = -\frac{4}{3}$$

## UNDERSTANDING THE MAIN IDEAS

You can solve a quadratic equation that is in the form $ax^2 + bx + c = 0$ by using the quadratic formula.

### Example 1

Solve each equation using the quadratic formula.

**a.** $3x^2 + 10x + 3 = 0$ **b.** $2x^2 - 10x + 1 = 0$

### ■ Solution ■

**a.** In the equation $3x^2 + 10x + 3 = 0$, $a = 3$, $b = 10$, and $c = 3$.

$x = \dfrac{-b \pm \sqrt{b^2 - 4ac}}{2a}$ ← Substitute 10 for $b$, 3 for $a$, and 3 for $c$.

$= \dfrac{-10 \pm \sqrt{10^2 - 4(3)(3)}}{2(3)}$

$= \dfrac{-10 \pm \sqrt{100 - 36}}{6}$

$= \dfrac{-10 \pm \sqrt{64}}{6}$

$= \dfrac{-10 \pm 8}{6}$

There are two exact solutions.

$x = \dfrac{-10 + 8}{6}$ or $x = \dfrac{-10 - 8}{6}$

$= -\dfrac{1}{3}$ $= -3$

The solutions are $-\dfrac{1}{3}$ and $-3$.

**b.** In the equation $2x^2 - 10x + 1 = 0$, $a = 2$, $b = -10$, and $c = 1$.

$x = \dfrac{-b \pm \sqrt{b^2 - 4ac}}{2a}$ ← Substitute $-10$ for $b$, 2 for $a$, and 1 for $c$.

$= \dfrac{-(-10) \pm \sqrt{(-10)^2 - 4(2)(1)}}{2(2)}$

$= \dfrac{10 \pm \sqrt{100 - 8}}{4}$

$= \dfrac{10 \pm \sqrt{92}}{4}$

Since 92 is not a perfect square, there are not two exact solutions. The solutions can be given as decimal approximations.

$x \approx \dfrac{10 + 9.59}{4}$ or $x \approx \dfrac{10 - 9.59}{4}$

$\approx 4.90$ $\approx 0.10$

The solutions are about 4.9 and about 0.1.

**Solve each equation using the quadratic formula.**

1. $x^2 + 2x + 1 = 0$
2. $x^2 - 5x + 6 = 0$
3. $2n^2 + 7n + 3 = 0$
4. $4x^2 + 11x - 3 = 0$
5. $-2y^2 - 3y + 2 = 0$
6. $3x^2 - 7x = 0$
7. $4x^2 = 5x + 6$
8. $4x^2 - 3x = 6$
9. $5x^2 = 2x + 4$

### Example 2

A firefighter is spraying the water from a fire hose through a window on the fifth floor of a building. The equation $h = -0.05d^2 + 2.8d + 5$ models the path of the water, where $h$ = the height (in feet) of the water above the ground and $d$ = the horizontal distance (in feet) of the water from the nozzle of the hose. If the window is 44 ft above the ground, how far from the building is the firefighter standing?

### ■ Solution ■

Replace $h$ in the equation with 44 and rewrite the equation in the form $ax^2 + bx + c = 0$.

$$44 = -0.05d^2 + 2.8d + 5$$
$$0 = -0.05d^2 + 2.8d - 39$$

Now solve this quadratic equation for $d$ using the quadratic formula.

$$x = \frac{-b \pm \sqrt{b^2 - 4ac}}{2a} \quad \leftarrow \text{Substitute 2.8 for } b, -0.05 \text{ for } a, \text{ and } -39 \text{ for } c.$$

$$= \frac{-2.8 \pm \sqrt{(2.8)^2 - 4(-0.05)(-39)}}{2(-0.05)}$$

$$= \frac{-2.8 \pm \sqrt{7.84 - 7.8}}{-0.1}$$

$$= \frac{-2.8 \pm \sqrt{0.04}}{-0.1}$$

$$= \frac{-2.8 \pm 0.2}{-0.1}$$

$x = \dfrac{-2.8 + 0.2}{-0.1}$ or $x = \dfrac{-2.8 - 0.2}{-0.1}$

$= 26 \qquad\qquad = 30$

The firefighter is standing either 26 ft or 30 ft from the building.

10. Refer to Example 2. How far away from the building should the firefighter be to get water into a window that is 34 ft above ground level?

11. **Mathematics Journal** In some real-world situations, only one of the solutions to a quadratic equation makes sense and the other solution is omitted. Explain why both solutions in Example 2 are included as answers. Do you think one of the solutions is a better answer? If so, which one and why.

## Spiral Review

**Tell whether each number is *rational* or *irrational*.** *(Section 6.2)*

**12.** $\sqrt{49}$     **13.** $\sqrt{188}$     **14.** $\sqrt{289}$

**Use substitution to solve each system of equations.** *(Section 7.2)*

**15.** $y = 3x + 1$     **16.** $y = -x - 1$     **17.** $x = 2y - 2$
$2x - y = -6$     $2x + y = 2$     $x - 3y = -6$

# Section 8.6 Using the Discriminant

**GOAL**

**Learn how to...**
- find the discriminant of a quadratic equation

**So you can...**
- find the number of solutions of a quadratic equation

*Application*

The path of a water arc is modeled by the equation $y = -0.005x^2 + x + 10$, where $x$ = the horizontal distance and $y$ = the vertical distance traveled by the water.

### Terms to Know

**Discriminant** (p. 355)
the value of the expression $b^2 - 4ac$ in the quadratic formula

### Example / Illustration

If $2x^2 + 5x + 1 = 0$, then
$$b^2 - 4ac = 5^2 - 4(2)(1)$$
$$= 25 - 8$$
$$= 17$$

## UNDERSTANDING THE MAIN IDEAS

The value of the discriminant tells you how many real-number solutions there are to a quadratic equation.

When $b^2 - 4ac > 0$, there are **two** solutions.

When $b^2 - 4ac = 0$, there is **one** solution.

When $b^2 - 4ac < 0$, there are **no** solutions.

When a quadratic equation is graphed, the number of $x$-intercepts of the graph corresponds to the number of real-number solutions indicated by the value of the discriminant.

### Example 1

For each equation, how many solutions are there?

**a.** $2x^2 + 4x + 1 = 0$   **b.** $x^2 - 6x + 9 = 0$   **c.** $2x^2 + 1 = -2x$

### Solution

**a.** For the quadratic equation $2x^2 + 4x + 1 = 0$, the values of $a$, $b$, and $c$ are 2, 4, and 1, respectively.

$$b^2 - 4ac = 4^2 - 4(2)(1)$$
$$= 16 - 8$$
$$= 8$$

Since $b^2 - 4ac > 0$, there are two solutions of the equation.

**b.** The values of $a$, $b$, and $c$ are 1, –6, and 9, respectively.

$$b^2 - 4ac = (-6)^2 - 4(1)(9)$$
$$= 36 - 36$$
$$= 0$$

Since $b^2 - 4ac = 0$, there is one solution of the equation.

**c.** Be sure to rewrite the equation in the form $ax^2 + bx + c = 0$ before identifying the values of $a$, $b$, and $c$: $2x^2 + 2x + 1 = 0$.

The values of $a$, $b$, and $c$ are 2, 2, and 1, respectively.

$$b^2 - 4ac = 2^2 - 4(2)(1)$$
$$= 4 - 8$$
$$= -4$$

Since $b^2 - 4ac < 0$, there are no solutions of the equation.

**For each equation, find the number of solutions.**

1. $x^2 - 3x - 18 = 0$
2. $n^2 - 2n + 1 = 0$
3. $3x^2 + 12 = 0$
4. $3y^2 - 10y + 3 = 0$
5. $2m^2 - 5m = 0$
6. $8y^2 + 5 = 10y$
7. $5x^2 + 4x - 1 = 0$
8. $3n^2 - 9 = 0$
9. $4z^2 - 4z = -1$

**For each equation, find the number of solutions. Then solve using the quadratic formula.**

10. $-m^2 + 3m = 2$
11. $3x^2 - 4x = 6$
12. $3n^2 + 8n + 7 = 0$
13. $4a^2 - 20a = -25$
14. $3b^2 = 4b + 5$
15. $9x^2 + 12x + 4 = 0$

### Example 2

The path of a water arc in Pleasantville is modeled by the equation $y = -0.005x^2 + x + 10$, where $x$ = the horizontal distance and $y$ = the vertical distance traveled by the water. How many solutions are there of the equation when $y = 0$? when $y = 60$?

> **Solution**
>
> When $y = 0$, the equation is $-0.005x^2 + x + 10 = 0$. The value of the discriminant for this equation is:
>
> $$b^2 - 4ac = (1)^2 - 4(-0.005)(10) \quad \leftarrow b = 1, a = -0.005, c = 10$$
> $$= 1 + 0.2$$
> $$= 1.2$$
>
> Since the value of the discriminant is positive, there are two solutions.
>
> When $y = 60$, the equation is $-0.005x^2 + x + 10 = 60$, or $-0.005x^2 + x - 50 = 0$. The value of the discriminant for this equation is:
>
> $$b^2 - 4ac = (1)^2 - 4(-0.005)(-50) \quad \leftarrow b = 1, a = -0.005, c = -50$$
> $$= 1 - 1$$
> $$= 0$$
>
> Since the value of the discriminant is zero, there is one solution.

**16.** How many solutions does the equation of the water arc have when $y = 10$? What are the solutions?

**17.** Explain why there is only one solution to the equation when $y = 60$.

## Spiral Review

**Solve each system of equations.** *(Section 7.4)*

**18.** $3x + 5y = -14$
$2x + y = 0$

**19.** $x - 2y = -13$
$3x + y = 3$

**Solve each equation.** *(Section 8.3)*

**20.** $2x^2 - 4 = 0$

**21.** $5x^2 - 80 = 0$

**22.** $\frac{3}{4}m^2 - 23 = 25$

# Chapter 8 Review

**CHAPTER CHECK-UP**

Complete these exercises for a review of Chapter 8. If you have difficulty with a particular problem, review the indicated section.

For Exercises 1–3, graph each equation. Tell whether each equation is *linear* or *nonlinear*. *(Section 8.1)*

**1.** $y = 2x^2$  **2.** $y = |x|$  **3.** $y = 2x - 2$

**4.** The Drama Club's ticket committee is ready to make a decision on ticket prices for this year's spring show. Their estimates, based on past experience, show the following relationship between ticket price, number of tickets sold, and total income.

| Ticket price | Number of tickets sold | Total income |
|---|---|---|
| $2.50 | 350 | $875 |
| $3.00 | 300 | $900 |
| $3.50 | 250 | $875 |
| $4.00 | 200 | $800 |
| $4.50 | 150 | $675 |

Do you think the relationship between ticket price and total income is *linear* or *nonlinear*? Explain your answer. *(Section 8.1)*

For Exercises 5–7:
  a. Predict how each graph will compare with the graph of $y = x^2$.
  b. Sketch the graph. *(Section 8.2)*

**5.** $y = 3x^2$  **6.** $y = -\frac{1}{3}x^2$  **7.** $y = 0.25x^2$

For Exercises 8–10, solve each equation algebraically or by using a graph. *(Section 8.3)*

**8.** $28 = 4n^2$  **9.** $-4x^2 - 15 = 5$  **10.** $\frac{2}{3}a^2 - 17 = 7$

**11.** A ball dropped from a window 50 ft above the ground falls a distance $d = 16t^2$, where $d$ = the distance in feet and $t$ = the time in seconds. How long does it take the ball to hit the ground? *(Section 8.3)*

Use the related graph to solve each equation. If an equation has no solution, write *no solution*. *(Section 8.4)*

**12.** $2x^2 = 9x - 4$  **13.** $x^2 - 12x + 36 = 0$  **14.** $3x^2 - 4x + 12 = 0$

**Solve each equation using the quadratic formula.** *(Section 8.5)*

**15.** $x^2 - 10x + 8 = 0$
**16.** $3y^2 - 27y = 0$
**17.** $2x^2 + 8x = -9$

**For each equation in Exercises 18–20, find the number of solutions. Then solve using the quadratic formula.** *(Section 8.6)*

**18.** $x^2 - 4x + 4 = 0$
**19.** $-m^2 + 3m + 5 = 7$
**20.** $y^2 + 13y + 12 = 0$

**21.** The path of a water arc in Pleasantville is modeled by the equation $y = -0.005x^2 + x + 10$, where $x =$ the horizontal distance and $y =$ the vertical distance traveled by the water. How many solutions are there of the equation when $y = 0$? How many of these solutions make sense in this situation? Give the solution(s). *(Section 8.6)*

## SPIRAL REVIEW  Chapters 1–8

**Solve each system of equations.**

**1.** $3x + 5y = 10$
   $2x + y = 2$
**2.** $x - 2y = 3$
   $3x + y = -5$
**3.** $y = -x - 1$
   $2x + y = -4$

**Solve each equation.**

**4.** $2x^2 - 4x + 8 = 0$
**5.** $5x^2 + 8x + 2 = 0$

**Simplify each expression.**

**6.** $\sqrt{(-5)^2 - 4(1)(-3)}$
**7.** $\sqrt{(-2)^2 - 4(-7)(2)}$

**Write each repeating decimal as a fraction in lowest terms.**

**8.** $0.0666\ldots$
**9.** $0.888\ldots$
**10.** $0.8333\ldots$

**Simplify each expression.**

**11.** $\sqrt{12}$
**12.** $\sqrt{\dfrac{1}{9}}$
**13.** $\sqrt{5} + \sqrt{125}$

**Find each product.**

**14.** $(x + 7)^2$
**15.** $(5x + 1)(5x - 1)$
**16.** $(3n + 1)^2$
**17.** $2a(4a - 17)$
**18.** $(y + 7)(y - 4)$
**19.** $(8 + \sqrt{3})(5 - \sqrt{3})$

**Graph each equation.**

**20.** $2x - 3y = 6$
**21.** $2x - 7y = 14$
**22.** $5y - 3x = 9$

# Section 9.1 Exponents and Powers

**GOAL**

**Learn how to . . .**
- show multiplication with exponents
- evaluate expressions that involve powers

**So you can . . .**
- find the volumes of buildings and figures

### Application

The transmitting distance of a cellular phone is the maximum distance from a transmitting tower within which the phone can operate. This distance depends on the atmospheric conditions and the terrain. For example, it is usually not possible to make a phone call in a valley. Most cellular phones have an estimated transmitting distance of 20 mi. The signal area covered by a tower is a circle and this area can be represented by the expression below which involves a power.

$$A \approx 3.14(20^2)$$

### Terms to Know

| Terms to Know | Example / Illustration |
|---|---|
| **Base (p. 370)** <br> the number being used as a factor | In the expression $3^2$, 3 is the base. |
| **Exponent (p. 370)** <br> the number that tells you how many times to multiply by the base | In the expression $4^5$, 5 is the exponent. |
| **Power (p. 370)** <br> an expression that includes a base and an exponent | The expression $5^3$ is read "five to the third power." |

### UNDERSTANDING THE MAIN IDEAS

An exponent is written above and to the right of a base. Expressions that include exponents are called powers. Any number can be expressed as a power. The exponent tells how many times to use the base as a factor.

### Example 1

Rewrite each product using exponents.

a. $3 \cdot 3 \cdot 3 \cdot 3$

b. $\left(-\frac{3}{8}\right)\left(-\frac{3}{8}\right)\left(-\frac{3}{8}\right)\left(-\frac{3}{8}\right)\left(-\frac{3}{8}\right)$

c. $-(a \cdot a \cdot a \cdot a \cdot a \cdot a)$

d. $n(4r \cdot 4r \cdot 4r \cdot 4r)$

### Solution

First identify the base; this is the repeated factor in the expression. Then determine the exponent; this is the number of times that the base is used as a factor.

a. The number 3 is used as a factor 4 times, so 3 is the base and 4 is the exponent.

$$3 \cdot 3 \cdot 3 \cdot 3 = 3^4$$

b. The fraction $-\frac{3}{8}$ is used as a factor 5 times, so $-\frac{3}{8}$ is the base and 5 is the exponent. Use parentheses to show that $-\frac{3}{8}$ (and not just $\frac{3}{8}$) is used as a factor 5 times.

$$\left(-\frac{3}{8}\right)\left(-\frac{3}{8}\right)\left(-\frac{3}{8}\right)\left(-\frac{3}{8}\right)\left(-\frac{3}{8}\right) = \left(-\frac{3}{8}\right)^5$$

c. The variable $a$ is used as a factor 6 times, so the base is $a$ and the exponent is 6.

$-(a \cdot a \cdot a \cdot a \cdot a \cdot a) = -(a^6)$ or $-a^6$   ← Note: $-a^6 \neq (-a)^6$

d. Use parentheses to show that both 4 and $r$ are being multiplied 4 times, but $n$ is used as a factor only once.

$$n(4r \cdot 4r \cdot 4r \cdot 4r) = n(4r)^4$$

**Rewrite each product using exponents.**

1. $5 \cdot 5 \cdot 5 \cdot 5 \cdot 5$

2. $14 \cdot 14 \cdot 14$

3. $\left(\frac{2}{5}\right)\left(\frac{2}{5}\right)\left(\frac{2}{5}\right)\left(\frac{2}{5}\right)\left(\frac{2}{5}\right)\left(\frac{2}{5}\right)$

4. $7 \cdot 7 \cdot 7(a+b)$

5. $d(-4e)(-4e)(-4e)$

6. $-(ab \cdot ab \cdot ab \cdot ab \cdot ab)$

### Example 2

Evaluate each power.

a. $5^4$

b. $\left(-\frac{2}{3}\right)^5$

c. $(542.7)^1$

### Solution

a. $5^4 = 5 \cdot 5 \cdot 5 \cdot 5$
$= 625$

b. $\left(-\dfrac{2}{3}\right)^5 = -\dfrac{2}{3} \cdot -\dfrac{2}{3} \cdot -\dfrac{2}{3} \cdot -\dfrac{2}{3} \cdot -\dfrac{2}{3}$
$= \dfrac{(-2)(-2)(-2)(-2)(-2)}{(3)(3)(3)(3)(3)}$
$= \dfrac{-32}{243}$, or $-\dfrac{32}{243}$

c. An exponent of 1 means that the base is used as a factor only once.
$(542.7)^1 = 542.7$

**Evaluate each power.**

7. $6^3$
8. $3^5$
9. $10^4$
10. $\dfrac{1}{4^6}$
11. $6.32^1$
12. $\left(-\dfrac{5}{8}\right)^2$

## Powers and order of operations

Evaluating powers is another form of multiplication. In the order of operations, evaluating powers is done before other multiplication and division.

*Order of Operations*

1. Simplify expressions inside parentheses.
2. Evaluate powers.
3. Do multiplication and division from left to right.
4. Do addition and subtraction from left to right.

### Example 3

Evaluate the expression $x^3(y^2 - 5) + y$ for $x = -2$ and $y = 4$.

### Solution

Substitute $-2$ for $x$ and $4$ for $y$.

$x^3(y^2 - 5) + y = (-2)^3(4^2 - 5) + 4$
$= (-2)^3(16 - 5) + 4$
$= (-2)^3(11) + 4$
$= (-8)(11) + 4$
$= -88 + 4$
$= -84$

**For Exercises 13–16, evaluate each expression when $x = 3$ and $y = -2$.**

**13.** $2x^2 - 7$  **14.** $x^3 + y^5$  **15.** $xy^4$  **16.** $(xy)^4$

**17. Open-ended Problem** Choose any base greater than 0 and less than 1. Write and evaluate several powers using your base. Describe the result.

**18.** Find the volume of a cube with sides 14.5 in. long. Recall that the volume of a cube is given by the formula $V = s^3$.

**19. Writing** Evaluate $-8$ to the first through fifth powers. Give a rule that can be used to determine when the power of a negative number will be positive and when it will be negative.

## Spiral Review

**Describe a situation that can be represented by each expression. Then evaluate the expression when $x = 5$ and $y = -4$.** *(Section 1.6)*

**20.** $x + \frac{1}{3}y$  **21.** $0.4(3x + 2)$  **22.** $y - \frac{2}{11}$

**Find each product. Then check the product by evaluating both sides of your equation when $m = 3$ and $n = -5$.** *(Section 6.5)*

**23.** $(m + 6)^2$  **24.** $(n + 3)(n - 3)$  **25.** $(6m - 5)^2$

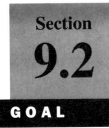

# Section 9.2 Exponential Growth

**GOAL**

**Learn how to . . .**
- interpret and evaluate exponential functions
- model real-world situations with exponential functions

**So you can . . .**
- make predictions about things that grow, like population

## Application

Certificates of deposit (CDs) generally offer higher rates of interest than regular savings accounts. The depositor agrees to leave the money in the bank for a set period of time and the bank guarantees a rate of interest for that time period. As the deposit remains in the bank earning interest it increases exponentially. The table below shows the CD interest rate at a certain bank.

| CD Rates (Minimum Deposit $500) ||
|---|---|
| Term | Annual percentage rate |
| 6 months | 5.38 |
| 1 year | 5.54 |
| 15 months | 5.59 |
| 18 months | 5.59 |
| 2 years | 5.80 |

## Terms to Know / Example / Illustration

| Terms to Know | Example / Illustration |
|---|---|
| **Exponential growth (p. 377)** an increase by a fixed percent each time period | The increase in the amount of money in a CD account over the term of the deposit is an example of exponential growth. It can be modeled by the equation $y = a(1 + r)^x$. |
| **Growth rate (p. 377)** the percent of growth expressed as a decimal (The growth rate is represented by $r$ in the exponential growth equation.) | The growth rate for the 1-year CD shown in the table in the Application is 0.0554. |
| **Growth factor (p. 377)** the quantity $1 + r$, where $r =$ the growth rate | The growth factor for the 2-year CD in the Application is $1 + 0.058$, or 1.058. |

**Study Guide,** ALGEBRA 1: EXPLORATIONS AND APPLICATIONS
Copyright © McDougal Littell Inc. All rights reserved.

# UNDERSTANDING THE MAIN IDEAS

## Exponential equations

A number increases exponentially when it increases by a fixed percent per time period. The time period can be any unit of time: seconds, days, month, years, and so on. The equation for modeling exponential growth is $y = a(1 + r)^x$. In the equation, the constant $a$ represents the initial amount before any growth occurs, the constant $r$ represents the growth rate, and $x$ represents the number of time periods.

A graph is another way to show exponential growth. The graph will be an upward curve.

### Example 1

Find the value of $y$ for the equation $y = 22(2.4)^x$ when $x = 5$.

### Solution

Substitute 5 for $x$.

$y = 22(2.4)^5$ ← Follow the order of operations.

$= 22 \cdot 79.62624$

$= 1751.77728$

**Find the value of $y$ when $x = 3$.**

1. $y = 3(2.75)^x$
2. $y = 25(1.23)^x$
3. $y = 6.2^x$
4. $y = 103(6)^x$
5. $y = 16.8(4.5)^x$
6. $y = 82(1.2)^x$

### Example 2

A newly-hatched channel catfish typically weighs about 0.3 g. This weight increases by an average of 10% per day for the first six weeks of the catfish's life.

a. Write an equation that gives the weight $y$ of a catfish in grams $x$ days after hatching.

b. Estimate a catfish's weight 21 days after hatching.

> **Solution**
>
> **a.** The catfish gains weight exponentially. Use the equation $y = a(1 + r)^x$, where $a$ = the original amount before growth, $r$ = the percent of growth (as a decimal), and $x$ = the number of units of time (in this case, days). So, $a = 0.3$ and $r = 10$.
>
> $y = a(1 + r)^x$ ← Substitute 0.3 for $a$ and 0.1 for $r$.
>
> $y = 0.3(1 + 0.1)^x$
>
> The equation is $y = 0.3(1 + 0.1)^x$, or $y = 0.3(1.1)^x$.
>
> **b.** Substitute 21 for $x$ in the equation from part (a).
>
> $y = 0.3(1.1)^x$
>
> $\phantom{y} = 0.3(1.1)^{21}$ ← Use a calculator.
>
> $\phantom{y} = 0.3 \cdot 7.4$ ← Round to the nearest hundredth.
>
> $\phantom{y} \approx 2.2$ ← Round to the nearest tenth.
>
> The catfish weighs about 2.2 g after 21 days.

**Suppose that today you invested $750 in a 4-year certificate of deposit paying 6% interest.**

7. Write an equation that gives the value $y$ of the CD (in dollars) $x$ years from now.

8. How much will the account be worth after 2 years? after 4 years?

**In 1985, the national debt of the United States was $1823.1 billion. The debt has increased at an average rate of 11.5% since that year.**

9. Write an equation that gives the amount $y$ of the national debt of the United States in billions of dollars $x$ years after 1985.

10. Estimate the national debt at the end of the year 2000.

## Spiral Review

**Solve each system of equations by adding or subtracting.** *(Section 7.3)*

11. $x + y = 1$
    $x - y = 15$

12. $6m - 3n = 11$
    $6m + n = 3$

13. $12a - 3b = 6$
    $-12a + 5b = 2$

14. $3r + 2s = 4$
    $5r + 2s = 6$

15. $5x + 3y = 11$
    $5x - 4y = -3$

16. $2p + 5q = 9$
    $-2p - 3q = 5$

**Apply the quadratic formula to solve each equation.** *(Section 8.5)*

17. $x^2 - 3x + 2 = 0$

18. $y^2 + 2y - 3 = 0$

19. $x^2 + 7x - 2 = 0$

20. $2x^2 - 4x + 1 = 0$

21. $3y^2 + 4y = 1$

22. $3 + 2x^2 = 3x^2 + 5x$

Study Guide, **ALGEBRA 1: EXPLORATIONS AND APPLICATIONS**

# Section 9.3 Exponential Decay

**GOAL**

**Learn how to...**
- use equations to model real-world decay situations

**So you can...**
- make predictions about quantities that decrease

## Application

The owner of an office supply store is buying a $12,000 car to be used for deliveries. As soon as the owner drives the car off the car dealer's lot its value begins to decrease at a rate of 40% per year. This decrease in value is called *depreciation* and the amount is given by the formula $V = 12,000(1 - 0.4)^x$, where $x$ = the number of years since the car was purchased. Depreciation is an example of exponential decay.

### Terms to Know / Example / Illustration

| Terms to Know | Example / Illustration |
|---|---|
| **Exponential decay** (p. 384) — a quantity that decreases regularly by a fixed percent | An annual decrease in the population of a country or city is an example of exponential decay. It can be modeled by the equation $y = a(1 - r)^x$. |
| **Rate of decrease** (p. 384) — the percent of decrease expressed as a decimal (The rate of decrease is represented by $r$ in the exponential decay formula.) | In the equation $V = 12,000(1 - 0.4)^x$, 0.4 is the rate of decrease. |
| **Decay factor** (p. 384) — the quantity $1 - r$, where $r$ = the rate of decrease | In the equation $V = 12,000(1 - 0.4)^x$, the decay factor is $1 - 0.4$, or 0.6. |

## UNDERSTANDING THE MAIN IDEAS

Exponential decay is the opposite of exponential growth. On a graph it is a downward curve. It represents systematic, continuous negative change. The rate of decrease may be an average percent over a period of time.

### Example 1

Find the value of $y$ for the equation $y = 278(0.93)^x$ when $x = 6$.

### Solution

Substitute 6 for $x$.

$y = 278(0.93)^6$     ← Use a calculator.

$\approx 278(0.65)$     ← Round to the nearest hundredth.

$\approx 180.7$

**Find the value of $y$ when $x = 5$.**

**1.** $y = 35(0.34)^x$          **2.** $y = 412.5(0.92)^x$          **3.** $y = 52.6(0.45)^x$

## Example 2

The amount $y$ of a substance decreases by 17% each hour after 3:00 P.M. At 3:00 P.M., there is a 100 g of the substance. How many grams of the substance remain 8 h later?

### Solution

The substance decays exponentially. Use the equation $y = a(1 - r)^x$, where $a$ = the original amount before decay, $r$ = the percent of decrease (as a decimal), and $x$ = the number of units of time (in this case, hours). So, $a = 100$, $r = 0.17$, and $x = 8$.

$y = a(1 - r)^x$     ← Substitute 100 for $a$, 0.17 for $r$, and 8 for $x$.

$= 100(1 - 0.17)^8$

$= 100(0.83)^8$     ← Use a calculator.

$\approx 100(0.225)$     ← Round to the nearest thousandth.

$\approx 22.5$

The amount of the substance 8 h later is about 22.5 g.

## Example 3

In 1990, the population of a city was 850,000. Each year since 1990, the population has been about 97% of what it was the previous year. If this rate of decrease continues, estimate the population of the city in the year 2000.

> **■ Solution ■**
>
> The population is decreasing exponentially, so use the equation $y = a(1 - r)^x$.
>
> The population in 1990 was 850,000, so $a = 850,000$. Since the amount after the decrease each year is 97% of the previous year's population, the rate of decrease is $100 - 97$, or 3%. So, $r = 0.03$. Since it is 10 years from 1990 to 2000, $x = 10$.
>
> $y = a(1 - r)^x$  ← Substitute 850,000 for $a$, 0.03 for $r$, and 10 for $x$.
> $= 850,000(1 - 0.03)^{10}$
> $= 850,000(0.97)^{10}$  ← Use a calculator.
> $\approx 850,000(0.737)$  ← Round to the nearest thousandth.
> $\approx 626,000$  ← Round to the nearest thousand.
>
> If this rate of decrease continues, the population of the city in the year 2000 will be about 626,000.

**A philanthropist plans to leave 50,000 shares of stock to a scholarship fund. The fund may use 5% of the stock each year for student scholarships.**

4. Write an equation that models the number of shares $y$ of stock remaining after $x$ years.

5. Find the number of shares of stock remaining after 35 years.

6. Will the stock fund ever run out? Explain your answer.

**Andres bought an $18,000 car. The car will depreciate at a rate of 20% per year.**

7. Write an equation that gives the value $y$ of the car after $x$ years.

8. Find the value of Andres' car after 4 years.

9. Andres took out a 5-year loan with a monthly payment of $393.62. If he trades in the car after 5 years and receives its full "depreciated" value, what will be the total amount he spent on the car?

**For Exercises 10 and 11, tell whether the graph represents exponential growth or exponential decay. How can you tell?**

10.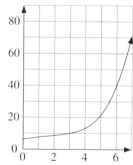

11.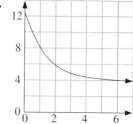

**12. Open-ended Problem** Write two equations, one that models the exponential growth of a quantity and one that models the exponential decay of a quantity. Identify the initial amount and give the growth or decay factor for each quantity.

**13. Mathematics Journal** Infrastructure is the system of roads, utilities, and public facilities that are needed for a community to operate. Describe how a city planner might use exponential growth or exponential decay to make decisions about changes in the infrastructure of a city.

## Spiral Review

**Simplify each expression.** *(Section 1.5)*

**14.** $7 + 3 - 12 \cdot \dfrac{2}{6}$

**15.** $\dfrac{2}{3} \cdot 18 \cdot (-7)$

**16.** $\dfrac{1}{4} \div \dfrac{1}{2} \cdot 6 \div (-3)$

**Identify the reciprocal you would use to solve each equation. Then solve.** *(Section 4.2)*

**17.** $\dfrac{3}{7}e = f$

**18.** $\dfrac{1}{2}a = -b$

**19.** $\dfrac{2}{5}m = 3n$

**20.** $-\dfrac{2}{3}r = s$

# Section 9.4 Zero and Negative Exponents

**GOAL**

**Learn how to . . .**
- evaluate powers with negative and zero exponents
- use negative exponents to express quotients

**So you can . . .**
- find past values
- use formulas to find information

## Application

Buying shares of stock in a company is considered to be an excellent long-term investment. Since 1945, stocks have increased in value an average of 11% per year. The Dow Jones Industrials Average tracks the price of 30 stocks. On one day in July, 1995, the Dow Jones Industrials Average was 4732.77. You can use a zero exponent to represent the average on this day and negative exponents to estimate the average before this day.

$$A = 4732.77(1 + 0.11)^0 \qquad A = 4732.77(1 + 0.11)^{-x}$$

## UNDERSTANDING THE MAIN IDEAS

Any nonzero number to the zero power is 1. That is, for any nonzero number $a$, $a^0 = 1$.

Also, for any nonzero number $a$ and any integer $n$, $a^{-n} = \dfrac{1}{a^n}$.

### Example 1

Simplify each expression.

**a.** $\left(\dfrac{1}{5}\right)^0$     **b.** $2^{-5}$     **c.** $\left(\dfrac{3}{4}\right)^{-2}$

### Solution

**a.** $\left(\dfrac{1}{5}\right)^0 = 1$   ← Any nonzero number to the zero power is 1.

**b.** $2^{-5} = \dfrac{1}{2^5} = \dfrac{1}{32}$

**c.** Remember that the reciprocal of $a$ is $\dfrac{1}{a}$. Since $1^n = 1$ for any value $n$, the rule for finding negative exponents could be stated as $a^{-n} = (\text{reciprocal of } a)^n$.

$\left(\dfrac{3}{4}\right)^{-2} = \left(\dfrac{4}{3}\right)^2$   ← The reciprocal of $\dfrac{3}{4}$ is $\dfrac{4}{3}$.

$\approx 1.8$   ← Use your calculator.

### Example 2

Rewrite each expression using only positive exponents.

a. $a^{-2}$　　b. $\left(\dfrac{a}{b}\right)^{-3}$　　c. $ab^2c^{-1}$

### ■ Solution ■

Use the rule "For any nonzero number $a$ and any integer $n$, $a^{-n} = \dfrac{1}{a^n}$."

a. $a^{-2} = \dfrac{1}{a^2}$

b. $\left(\dfrac{a}{b}\right)^{-3} = \left(\dfrac{b}{a}\right)^3$　← The reciprocal of $\dfrac{a}{b}$ is $\dfrac{b}{a}$.

c. $ab^2c^{-1} = ab^2 \cdot \dfrac{1}{c}$

　　　　$= \dfrac{ab^2}{c}$

**Simplify each expression.**

1. $6^0$
2. $2^{-8}$
3. $5^{-2}$
4. $3^{-5}$
5. $\left(\dfrac{2}{5}\right)^{-3}$
6. $\left(\dfrac{1}{4}\right)^0$
7. $\left(\dfrac{6}{7}\right)^{-4}$
8. $\left(\dfrac{4}{5}\right)^{-3}$

**Rewrite each expression using only positive exponents.**

9. $a^{-6}$
10. $z^{-3}$
11. $\left(\dfrac{s}{t}\right)^{-5}$
12. $\dfrac{u}{v^{-2}}$
13. $mn^3o^{-3}$
14. $(ab)^{-4}$

When you take out a loan to buy a car, you must repay a portion of the loan plus interest on the loan each month. The monthly payment $P$ required to repay a loan of $A$ dollars in $n$ monthly installments is given by the formula

$$P = \dfrac{A \times m}{1 - (1 + m)^{-n}}$$

where $m$ = the monthly interest rate expressed as a decimal. To compute the monthly interest rate, divide the annual interest expressed as a decimal by 12.

### Example 3

Find the monthly payment on a $9995 car if you finance the entire cost at 7.9% interest and repay the loan over 36 months.

**Study Guide,** ALGEBRA 1: EXPLORATIONS AND APPLICATIONS
Copyright © McDougal Littell Inc. All rights reserved.

> **Solution**
>
> First find the monthly interest rate.
>
> $m = \dfrac{0.079}{12}$
>
> $\phantom{m} \approx 0.0066$ ← Round to the nearest ten thousandth.
>
> Now use the formula $P = \dfrac{A \times m}{1 - (1 + m)^{-n}}$ with $A = 9995$ and $n = 36$.
>
> $P = \dfrac{A \times m}{1 - (1 + m)^{-n}}$ ← Substitute 9995 for $A$, 0.0066 for $m$, and 36 for $n$.
>
> $\phantom{P} = \dfrac{9995(0.0066)}{1 - (1 + 0.0066)^{-36}}$
>
> $\phantom{P} = \dfrac{9995(0.0066)}{1 - (1.0066)^{-36}}$ ← Use a calculator.
>
> $\phantom{P} \approx \dfrac{9995(0.0066)}{1 - 0.7891}$ ← Round to the nearest ten thousandth.
>
> $\phantom{P} \approx \dfrac{65.967}{0.2109}$
>
> $\phantom{P} \approx 312.79$ ← Round to the nearest hundredth.
>
> The monthly payment is $312.79.

**For Exercises 15–17, suppose that you take out a loan for the entire cost of the given vehicle. What is your monthly payment if you repay the loan over each time period?**

**15.** 24 monthly payments on a $13,995 car at 2.9% interest

**16.** 60 monthly payments on a $14,595 truck at 10.5% interest

**17.** 48 monthly payments on a $7995 car at 9% interest

### Spiral Review

**Solve each inequality.** *(Section 4.6)*

**18.** $3 + x \leq 6$

**19.** $4r \geq r - 2$

**20.** $n + 37 < 11 - n$

**21.** $\dfrac{5c + 2}{3} > -8$

**22.** $3 < 2x - 4 < 10$

**23.** $-2 < 4 - 3n \leq 15$

# Section 9.5 Working with Scientific Notation

**GOAL**

**Learn how to...**
- write numbers in scientific notation

**So you can...**
- conveniently express numbers

### Application

The lethal Ebola virus is a microorganism just 0.8 microns in length. This is 0.0000008 m, or in scientific notation, $8 \times 10^{-7}$ m. A microscope would need to magnify the virus 2500 times for it to be the size of the head of a pin.

| Terms to Know | Example / Illustration |
|---|---|
| **Scientific notation (p. 396)** a number expressed in the form $a \times 10^n$ where $1 \leq a < 10$ and $n$ is an integer | The estimated world population in 1994 was about 5,600,000,000 or $5.6 \times 10^9$ in scientific notation. |

## UNDERSTANDING THE MAIN IDEAS

Scientific notation is used to express very large and very small numbers. Negative exponents are used in the notation for the small (decimal) numbers.

### Example 1

The world's largest pizza was made in 1990. It had a diameter of 122 ft 8 in.

**a.** Find the area of this pizza to the nearest hundred thousand square inches.

**b.** Write the area using scientific notation.

### ■ Solution ■

**a.** First convert the diameter 122 ft 8 in. to inches.

$$122 \text{ ft } 8 \text{ in.} = [(122 \cdot 12) + 8] \text{ in.}$$
$$= (1464 + 8) \text{ in.}$$
$$= 1472 \text{ in.}$$

Now use the formula for the area of a circle, $A = \pi r^2$. *Note:* Recall that the radius, $r$, is half the diameter.

*(Solution continues on next page.)*

### Solution (continued)

$A = \pi r^2$

$\approx 3.14 \cdot (0.5 \cdot 1472)^2$

$\approx 3.14 \cdot 541{,}696$

$\approx 1{,}700{,}000$ ← Round to the nearest hundred thousand.

The area of the pizza was about 1,700,000 in.$^2$.

**b.** Write 1,700,000 in scientific notation. The number must be in the form $a \times 10^n$ where $1 \leq a < 10$ and $n$ is an integer. For 1,700,000, the value of $a$ is 1.7. The decimal point has moved 6 places to the left, meaning we have factored out $10^6$.

So, 1,700,000 in.$^2$ = $1.7 \times 10^6$ in.$^2$.

When factoring out powers of 10, positive exponents correspond to moving the decimal point to the left to get the value of $a$ and negative exponents correspond to moving the decimal point to the right to get the value of $a$.

### Example 2

Write each number in scientific notation.

**a.** 36,000,000  **b.** 0.0000000502

### Solution

**a.** For 36,000,000, the value of $a$ is 3.6. The decimal point moved 7 places to the left to obtain $a$.

$36{,}000{,}000 = 3.6 \times 10^7$

**b.** For 0.0000000502, the value of $a$ is 5.02. The decimal point moved 8 places to the right to obtain $a$.

$0.0000000502 = 5.02 \times 10^{-8}$

The process is reversed when a number written in scientific notation is rewritten in decimal notation.

### Example 3

Write each number in decimal notation.

**a.** $3.67 \times 10^{15}$  **b.** $1.2 \times 10^{-8}$

> ### ■ Solution ■
>
> Follow the order of operations. Simplify the power before you multiply.
>
> **a.** $3.67 \times 10^{15} = 3.67 \times 1{,}000{,}000{,}000{,}000{,}000$  ← *Note*: For $n > 0$, $10^n$ is a 1 followed by $n$ zeros.
> $= 3{,}670{,}000{,}000{,}000{,}000$
>
> Notice that the number of digits in the answer is the same as the number of digits in the power of 10.
>
> **b.** $1.2 \times 10^{-8} = 1.2 \times 0.00000001$  ← *Note*: For $n > 0$, $10^{-n}$ is a 1 preceeded by a decimal point and $(n-1)$ zeros.
> $= 0.000000012$
>
> Notice that the number of zeros in the answer is the same as the number of zeros in the power of 10.

**Write each number in decimal notation.**

1. $6 \times 10^3$
2. $8 \times 10^{-4}$
3. $5 \times 10^9$
4. $2.7 \times 10^8$
5. $5.13 \times 10^{11}$
6. $9.177 \times 10^{-15}$

**Write each number in scientific notation.**

7. 300,000
8. 0.00000007
9. 16,000,000
10. 1,540,000,000
11. 0.000000249
12. 0.00003002

**Write each number in scientific notation.**

13. The sun rotates around the center of the Milky Way galaxy once every 210,000,000 years.

14. The mass of Earth's oceans in metric tons is 1,350,000,000,000,000,000.

15. Subatomic particles (particles that are smaller than atoms) have had an important role in the development of microwave technology. The heaviest subatomic particle, called a *hadron*, has a lifetime of 0.00000000000000000000000083 s.

................
### Spiral Review

**Write each expression in simplest form.** *(Section 6.3)*

16. $3^2 \cdot 7^5$
17. $(4 \cdot 3)^7$
18. $\dfrac{4^2}{8^4}$

**Simplify each power.** *(Section 9.1)*

19. $(-8)^6$
20. $\left(\dfrac{4}{5}\right)^5$
21. $\left(7 \cdot \dfrac{5}{7}\right)^4$

# Section 9.6

## Exploring Powers

**GOAL**

**Learn how to...**
- multiply and divide expressions involving powers

**So you can...**
- perform calculations

### Application

Nuclear scientists use extremely small numbers in their work with atoms. For example, the weights of atoms range from $1.7 \times 10^{-24}$ g for hydrogen to $3.4 \times 10^{-22}$ g for plutonium. Finding how many times heavier a plutonium atom is than a hydrogen atom requires finding the quotient of two powers.

### Terms to Know | Example / Illustration

| Terms to Know | Example / Illustration |
|---|---|
| **Product of powers rule** (p. 402)<br>a rule for multiplying powers with like bases; in symbols, $a^m \cdot a^n = a^{(m+n)}$ for any nonzero number $a$ and any integers $m$ and $n$ | $5^3 \cdot 5^2 = 5^{(3+2)}$<br>$= 5^5$<br>$= 3125$ |
| **Quotient of powers rule** (p. 402)<br>a rule for dividing powers with like bases; in symbols, $\dfrac{a^m}{a^n} = a^{(m-n)}$ for any nonzero number $a$ and any integers $m$ and $n$ | $\dfrac{6^3}{6^1} = 6^{(3-1)}$<br>$= 6^2$<br>$= 36$ |

## UNDERSTANDING THE MAIN IDEAS

The rules for multiplying or dividing two powers that have the same base are the *product of powers rule* and the *quotient of powers rule*. The product of powers rule tells us to add the exponents when the base is the same. The quotient of powers rule tells us to subtract the exponents when the base is the same.

Product of Powers Rule          Quotient of Powers Rule

$a^m \cdot a^n = a^{(m+n)}$          $\dfrac{a^m}{a^n} = a^{(m-n)}$

### Example 1

Simplify each expression.

**a.** $2^3 \cdot 2^4$       **b.** $5^7 \cdot 5^{-4}$       **c.** $\dfrac{10^6}{10^9}$

*Study Guide,* ALGEBRA 1: EXPLORATIONS AND APPLICATIONS

### Solution

**a.** The bases are the same. Follow the product of powers rule: add the exponents.

$$2^3 \cdot 2^4 = 2^{(3+4)}$$
$$= 2^7$$
$$= 128$$

**b.** $5^7 \cdot 5^{-4} = 5^{(7+(-4))}$
$$= 5^3$$
$$= 125$$

**c.** The bases are the same. Follow the quotient of powers rule: subtract the exponents.

$$\frac{10^6}{10^9} = 10^{(6-9)}$$
$$= 10^{(6+(-9))}$$
$$= 10^{-3}$$
$$= \frac{1}{10^3} \quad \leftarrow \text{Write the expression using positive exponents.}$$
$$= \frac{1}{1000}$$

**Write each expression as a single power.**

1. $3^2 \cdot 3^5$
2. $4^8 \cdot 4^{-3}$
3. $5^9 \cdot 5^2$
4. $\dfrac{7^{-2}}{7^4}$
5. $\dfrac{8^7}{8^4}$
6. $\dfrac{11^2}{11^{-7}}$

### Example 2

Simplify each expression.

**a.** $a^2 \cdot a^3$      **b.** $(5d^4)(8d^6)$

### Solution

**a.** $a^2 \cdot a^3 = a^{(2+3)}$
$$= a^5$$

**b.** $(5d^4)(8d^6) = (5)(8)(d^4)(d^6)$    $\leftarrow$ Use the associative property.
$$= 40(d^4)(d^6) \quad \leftarrow \text{Use the product of power rule.}$$
$$= 40 \cdot d^{(4+6)}$$
$$= 40d^{10}$$

**Study Guide,** ALGEBRA 1: EXPLORATIONS AND APPLICATIONS
Copyright © McDougal Littell Inc. All rights reserved.

### Example 3

**Simplify each expression.**

a. $\dfrac{b^{-9}}{b^4}$

b. $\dfrac{r^8 s^3}{r^6 s}$

### ■ Solution ■

a. $\dfrac{b^{-9}}{b^4} = b^{(-9-4)}$ ← Use the quotient of power rule.

$= b^{(-9 + (-4))}$

$= b^{-13}$

$= \dfrac{1}{b^{13}}$

b. $\dfrac{r^8 s^3}{r^6 s} = \dfrac{r^8}{r^6} \cdot \dfrac{s^3}{s}$ ← Group powers with the same base.

$= r^{(8-6)} \cdot s^{(3-1)}$

$= r^2 s^2$

**Simplify each expression.**

7. $r^4 \cdot 4r^3$

8. $n^3 \cdot n^7$

9. $(6s^{-2})(4s^3)$

10. $\dfrac{x^5}{x^{-2}}$

11. $\dfrac{2a^{-4}}{8a^3}$

12. $\dfrac{m^3 n^5}{m^2 n^4}$

### Example 4

The unmanned spacecraft *Voyager I* has traveled billions of miles into space. The tremendous distance slows down the transmission of messages between Earth and the spacecraft. About how long would it take a message traveling at the speed of light (186,281 mi/s) to be relayed to Earth if the spacecraft was as far away as Pluto ($3.57 \times 10^9$ mi)?

> **■ Solution ■**
>
> To estimate the answer, write the speed of light in scientific notation.
>
> $$186{,}281 \approx 1.86 \times 10^5 \qquad \leftarrow \text{Round to the nearest thousand.}$$
>
> Now divide the distance by the speed.
>
> $$\frac{3.57 \times 10^9}{1.86 \times 10^5} = \frac{3.57}{1.86} \times \frac{10^9}{10^5} \qquad \leftarrow \text{Group powers with the same base.}$$
>
> $$\approx 1.9 \times 10^{(9-5)} \qquad \leftarrow \text{Use the quotient of powers rule.}$$
>
> $$\approx 1.9 \times 10^4$$
>
> $$\approx 19{,}000$$
>
> It would take about 19,000 s, or about 5 h 16 min 40 s, to transmit a message back to Earth.

**13.** The maximum speed attained by manned space flights is about $2.5 \times 10^4$ mi/h. The distance from Earth to Pluto is about $3.57 \times 10^9$ mi. If it were possible for a space ship to fly directly to Pluto from Earth, what is the least amount of flight time?

**14.** The average distance from Mars to the sun is about $1.42 \times 10^8$ mi. Pluto, the planet farthest from the sun, has an average distance of about $3.66 \times 10^9$ mi from the sun. About how many times closer is Mars to the sun than Pluto?

**15.** The Bureau of Labor Statistics predicts that in the year 2002 there will be $1.5 \times 10^7$ executive jobs in the United States and $2.6 \times 10^5$ jobs in social science and urban planning. How many times greater are your chances of finding an executive position than a position in the social sciences or urban planning?

........................
## Spiral Review

**Evaluate each power.** *(Section 9.4)*

**16.** $\left(\dfrac{3}{5}\right)^0$

**17.** $\left(-\dfrac{61}{5}\right)^0$

**18.** $\left(\dfrac{5}{6}\right)^{-3}$

**19.** $\left(-\dfrac{7}{8}\right)^{-4}$

**20.** $-\left(\dfrac{1}{6}\right)^9$

**21.** $\dfrac{(13.8)^0}{(2.6)^{-5}}$

# Section 9.7 Working with Powers

**GOAL**

**Learn how to...**
- find the power of products and quotients
- find the power of a power

**So you can...**
- use numbers written in scientific notation in formulas

## Application

Supersonic jets are capable of flying at speeds greater than the speed of sound (called Mach 1). A United States Air Force reconnaissance aircraft, the world's fastest supersonic jet, has flown at Mach 3 speed. Mach 3 is about $2.3 \times 10^3$ mi/h.

Kinetic energy is the energy a body has due to its motion and is computed using the formula below.

$$k = 0.5(mv^2)$$

mass  velocity

Finding the kinetic energy of the aircraft would require finding the power of a power.

### Terms to Know / Example / Illustration

| Terms to Know | Example / Illustration |
|---|---|
| **Power of a product rule (p. 408)** a rule for evaluating the power of a product; in symbols, for any nonzero numbers $a$ and $b$ and any integer $m$, $(ab)^m = a^m b^m$ | $(2 \cdot 8)^4 = 2^4 \cdot 8^4$ |
| **Power of a quotient rule (p. 408)** a rule for evaluating the power of a quotient; in symbols, for any nonzero numbers $a$ and $b$ and any integer $m$, $\left(\dfrac{a}{b}\right)^m = \dfrac{a^m}{b^m}$ | $\left(\dfrac{2}{3}\right)^6 = \dfrac{2^6}{3^6}$ |
| **Power of a power rule (p. 408)** a rule for evaluating the power of a power; in symbols, for any nonzero number $a$ and any integers $m$ and $n$, $(a^m)^n = a^{m \cdot n}$ | $(8^4)^3 = 8^{4 \cdot 3}$ $= 8^{12}$ |

## UNDERSTANDING THE MAIN IDEAS

Some rules have been written to evaluate the power of a quantity. A quantity is any expression inside parentheses such as a product, a quotient, or another power. Rules for evaluating the power of a quantity are helpful when it is necessary to substitute expressions with exponents into formulas.

### Example 1

Evaluate $(3 \times 10^4)^2$. Write the answer in scientific notation.

**Solution**

First use the power of a product rule, then use the power of a power rule.

$(3 \times 10^4)^2 = 3^2 \cdot (10^4)^2$ ← Use the power of a product rule.

$\qquad\qquad\quad = 3^2 \cdot 10^{4 \cdot 2}$ ← Use the power of a power rule.

$\qquad\qquad\quad = 3^2 \cdot 10^8$

$\qquad\qquad\quad = 9 \times 10^8$

**Evaluate each power. Write your answers in scientific notation.**

**1.** $(6 \times 10^4)^3$      **2.** $(2 \times 10^6)^5$      **3.** $(7 \times 10^3)^8$

### Example 2

Evaluate $\left(\dfrac{3}{7}\right)^4$.

**Solution**

$\left(\dfrac{3}{7}\right)^4 = \dfrac{3^4}{7^4}$ ← Use the power of a quotient rule.

$\qquad\quad = \dfrac{81}{2401}$

**Evaluate each power.**

**4.** $\left(\dfrac{1}{8}\right)^7$      **5.** $\left(\dfrac{2}{3}\right)^6$      **6.** $\left(\dfrac{4}{5}\right)^3$

### Example 3

Simplify each expression.

**a.** $(rs)^5$      **b.** $\left(\dfrac{a^2}{b^3}\right)^5$

> **Solution**
>
> **a.** $(rs)^5 = r^5 s^5$ ← Use the power of a product rule.
>
> **b.** $\left(\dfrac{a^2}{b^3}\right)^5 = \dfrac{(a^2)^5}{(b^3)^5}$ ← Use the power of a quotient rule.
>
> $= \dfrac{a^{2 \cdot 5}}{b^{3 \cdot 5}}$ ← Use the power of a power rule.
>
> $= \dfrac{a^{10}}{b^{15}}$

**For Exercises 7–12, simplify each expression.**

**7.** $(xy)^6$  **8.** $(4ab)^3$  **9.** $(r^3 s^2)^5$

**10.** $\left(\dfrac{b}{c}\right)^4$  **11.** $\left(\dfrac{s^2}{y^5}\right)^2$  **12.** $\left(\dfrac{c^3 d^2}{e^8}\right)^5$

**13.** One of the smallest holes ever made was created by the beam of an electron microscope. It had a radius of $1 \times 10^{-9}$ m. What is the area of this hole? Use 3.14 for $\pi$.

**14. Open-ended Problem** In 1992, the population of a city dropped from 100,000 to 97,000. The population was 94,090 in 1993 and 91,267 in 1994. Describe how you would use a power of a quotient to describe the population trend in the city.

**15. Mathematics Journal** Find $(2 \times 10^3)^2$. Develop a method for checking your answer. Write a description of your method.

·················
## Spiral Review

**Simplify each variable expression.** *(Section 1.7)*

**16.** $2 + 5(m + 1) - 3m$  **17.** $5m^2 + 2m^2$

**18.** $3 + 2(4x - 7)$  **19.** $3(4p - 20) + 2(8p - 35)$

**20.** $7(c^2 + c) + 8(c^2 - c)$  **21.** $\dfrac{1}{2}(6a + 3b) + \dfrac{1}{4}(8a - 4b)$

# Chapter 9 Review

**CHAPTER CHECK-UP**

Complete these exercises for a review of Chapter 9. If you have difficulty with a particular problem, review the indicated section.

**Evaluate each power.** *(Section 9.1)*

**1.** $3^5$  **2.** $(-4)^6$  **3.** $\left(\dfrac{1}{8}\right)^3$

**Evaluate each expression when $x = 3$ and $y = 7$.** *(Section 9.1)*

**4.** $xy^3$  **5.** $(x + y)^5$  **6.** $x^3 + 4y^2$

**The population of Mexico City is increasing at a rate of 3% per year. In 1991, the population was about 21,000,000.** *(Section 9.2)*

**7.** Write an equation giving the number $y$ of people living in Mexico City $x$ years after 1991.

**8.** Estimate the population of Mexico City in the year 1998.

**Miko is planning to buy a $16,500 car. The depreciation rate for this car is 40% annually.** *(Section 9.3)*

**9.** Write an equation giving the value $y$ of Miko's car in $x$ years.

**10.** Miko has the option of financing the car for 24 months, 36 months, 48 months, or 60 months. Find the value of the car at the end of each of these finance terms.

**Evaluate each power.** *(Section 9.4)*

**11.** $14^0$  **12.** $3.25^{-3}$  **13.** $\left(\dfrac{8}{9}\right)^{-2}$

**Evaluate each expression when $x = -5$, $y = 3$, and $z = 6$.** *(Section 9.4)*

**14.** $x^4 y^{-1}$  **15.** $x^5 y^3 z^0$  **16.** $\left(\dfrac{y}{z}\right)^{-5}$

**The value of a piece of property has increased at an average annual rate of 7%. In 1995, the property was assessed at $524,500. Use the function $y = 524,500(1.07)^{-x}$, where $x$ = the number of years before 1995, to estimate the value of the property for each year.** *(Section 9.4)*

**17.** 1989  **18.** 1992  **19.** 1994

**Write each number in decimal notation.** *(Section 9.5)*

**20.** $3.764 \times 10^8$  **21.** $4.06 \times 10^{11}$  **22.** $8.1 \times 10^{-5}$

**Write each number in scientific notation.** *(Section 9.5)*

**23.** 0.0000007092  **24.** 34,000,000,000  **25.** 0.003

**Simplify each expression.** *(Sections 9.6 and 9.7)*

26. $a^3 \cdot a^8$

27. $\dfrac{x^{21}}{x^{18}}$

28. $(rs^2)^4$

29. $\left(\dfrac{5x^2}{3x^{-3}}\right)^2$

## SPIRAL REVIEW    Chapters 1–9

**Simplify each variable expression.**

1. $4r + 9r$

2. $13t^2 - 8t^2 + 7$

3. $2y^3 + 7y + 5y^3 - 3y$

**An artist is creating a square wall hanging for an art gallery opening. The length of each side of the wall hanging is (2s + 3) ft.**

4. What is the minimum length of trim material required to make edging for the wall hanging?

5. What is the area of the wall hanging?

6. The walls in the room where the wall hanging will be displayed are 10 ft high. The artist wants to leave at least 18 in. of space above and below her artwork. Write an inequality for the value of $s$.

**Simplify each expression.**

7. $r^3 \cdot r^8$

8. $\dfrac{n^{-6}}{n^4}$

9. $\left(\dfrac{x^{-7}y^2}{z^3}\right)^4$

**Solve each system of equations.**

10. $y = 3x - 4$
    $y = x + 3$

11. $x + 2y = 5$
    $-x + y = 3$

12. $3x - 5y = 7$
    $3x - y = 2$

**Solve each equation using the quadratic formula.**

13. $x^2 + 4x - 6 = 0$

14. $2x^2 - 3x - 2 = 0$

15. $x^2 - 5x + 1 = 0$

# Section 10.1

## Adding and Subtracting Polynomials

**GOAL**

**Learn how to . . .**
- recognize and classify polynomials
- add and subtract polynomials

**So you can . . .**
- use polynomials to model real-world situations

### Application

The total area of this figure can be expressed as $x^2 + xy$.

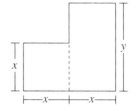

### Terms to Know

| Terms to Know | Example / Illustration |
|---|---|
| **Polynomial (p. 423)** an expression that can be written as a monomial or as a sum of monomials with exponents that are whole numbers | $10y + 10y^2 + 10y^3$ |
| **Standard form of a polynomial (p. 423)** a polynomial that is simplified, and with the exponents of the terms in order from largest to smallest | The polynomial $10y^3 + 10y^2 + 10y$ is in standard form. |
| **Degree of a term (p. 423)** the exponent of the variable (For a term with more than one variable, the degree is the sum of the exponents.) | The degree of the term $10y^2$ is 2. |
| **Degree of a polynomial (p. 423)** the largest degree of all the terms | The degree of the polynomial $10y^3 + 10y^2 + 10y$ is 3. |
| **Linear polynomial (p. 423)** a polynomial of degree 1 | $5a - 3b$ |
| **Quadratic polynomial (p. 423)** a polynomial of degree 2 | $3x^2 + 2x - 6$ |
| **Cubic polynomial (p. 423)** a polynomial of degree 3 | $15a^3 + 10a^2 - 5$ |

Study Guide, ALGEBRA 1: EXPLORATIONS AND APPLICATIONS
Copyright © McDougal Littell Inc. All rights reserved.

## Understanding the Main Ideas

To add or subtract polynomials, add or subtract the like terms.

### Example 1

Which of the following is a quadratic trinomial?

**A.** $x + y + 3$  **B.** $x^2 + x$  **C.** $2x^2 - x + 1$

**Solution**

Choice A has three terms and is degree 1; it is a linear trinomial.
Choice B has two terms and is degree 2; is a quadratic binomial.
Choice C has three terms and is degree 2; it is a quadratic trinomial.

Choice C is correct.

**Match each phrase with an algebraic expression.**

1. linear trinomial
2. quadratic binomial
3. cubic trinomial
4. not a polynomial

**A.** $2x^2 - x$
**B.** $\dfrac{1}{x^2}$
**C.** $x - 4y + 6$
**D.** $5y^3 + 3y^2 + 2$

### Example 2

Add $x^3 - 2x^2 + x - 4$ and $6x^2 - 3x - 5$.

**Solution**

Group the like terms.

$(x^3 - 2x^2 + x - 4) + (6x^2 - 3x - 5) = x^3 + (-2x^2 + 6x^2) + (x - 3x) + (-4 - 5)$
$= x^3 + 4x^2 - 2x - 9$

**Add. Give the degree of each sum.**

5. $(3c^2 + 2c - 4) + (c^2 - 7c + 3)$
6. $(4n^2 - n - 7) + (3n^3 + 2n^2 + n)$
7. $(m^2 + 4m + 3) + (-3m - 8)$
8. $(2x^2 - 7) + (-2x^2 + 2x)$

**Add.**

9. $(x + 2y + 4) + (-3x - y)$
10. $(4a^2 + ab + b^2) + (a^2 - b^2)$
11. $(3y^3 + y^2 + 9) + (-5y^2 - 6y + 7)$
12. $(g^3 + 8g - 2) + (3g^2 - 5g + 1)$

You can subtract two polynomials by adding the opposite of the second polynomial to the first. This process is similar to the method you learned for subtracting two integers: $-3 - 8 = -3 + (-8)$.

> **Example 3**
>
> Subtract $(n^3 - 5n^2 + n - 1) - (2n^2 - n - 3)$.
>
> ■ **Solution** ■
>
> Change each term of the polynomial being subtracted to its opposite and then add.
>
> $(n^3 - 5n^2 + n - 1) - (2n^2 - n - 3) = (n^3 - 5n^2 + n - 1) + (-2n^2 + n + 3)$
> $= n^3 + (-5n^2 - 2n^2) + (n + n) + (-1 + 3)$
> $= n^3 - 7n^2 + 2n + 2$

**Subtract.**

13. $(5n^2 + n + 8) - (3n^2 - 2n + 6)$
14. $(7m^2 - 5m - 2) - (-m^2 - 5m + 1)$
15. $(8x^2 - 2x + 7) - (4x^2 + x - 1)$
16. $(3y^2 - 6y + 10) - (-5y + 4)$
17. $(a^2 + 8) - (3a^2 - a + 1)$
18. $(3b - 5c + 7) - (b + c - 3)$
19. $(x^3 + 4x^2 + 6x - 3) - (5x^2 - x)$
20. $(9m^2 - n^2) - (7m^2 - 12mn + 3n^2)$

................
**Spiral Review**

**Find each product.** *(Section 6.4)*

21. $(y - 2)(y + 2)$
22. $(3x - 2)(x + 3)$
23. $(5x - 2)(2x - 5)$

**Estimate the solutions of each equation using the given graph.** *(Section 8.3)*

24. $x^2 - 3 = 0$
25. $-3x^2 + \dfrac{3}{2} = 0$
26. $\dfrac{1}{4}x^2 - 1 = 0$

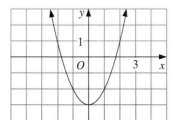
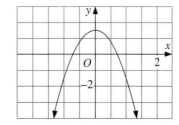
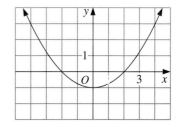

# Section 10.2 Multiplying Polynomials

**GOAL**

Learn how to . . .

- use the FOIL method to multiply binomials
- multiply polynomials

So you can . . .

- use polynomials to model lengths, areas, and volumes

## Application

The figure below shows Mr. McGregor's design for a vegetable garden. He wants each individual vegetable bed to be a square of side $x$ ft, and the walkways around the beds to be $y$ ft wide. Finding the area of the entire vegetable garden involves multiplying two polynomials.

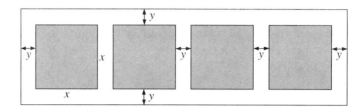

## UNDERSTANDING THE MAIN IDEAS

When you multiply two polynomials, each term in the first polynomial must be multiplied by each term in the second polynomial using the distributive property. Then the like terms are added together.

A second method, called the FOIL method, can be used to multiply two binomials. *FOIL* stands for "First terms of the binomials," "Outer terms of the expressions," "Inner terms of the expressions," and "Last terms of the binomials."

### Example 1

Multiply $(2x - 3)(3x + 5)$.

### Solution

**Method 1:** Use the distributive property.

$(2x - 3)(3x + 5) = 2x(3x + 5) - 3(3x + 5)$  ← Be sure to distribute the negative sign with the 3.

$\phantom{(2x - 3)(3x + 5)} = 6x^2 + 10x - 9x - 15$

$\phantom{(2x - 3)(3x + 5)} = 6x^2 + x - 15$

*(Solution continues on next page.)*

## Solution (continued)

*Method 2*: Use the FOIL method.

$$(2x - 3)(3x + 5) = 2x(3x) + 2x(5) + (-3)(3x) + (-3)(5)$$
$$= 6x^2 + 10x - 9x - 15$$
$$= 6x^2 + x - 15$$

(First, Last, Inner, Outer)

**Multiply.**

1. $(x + 2)(x + 7)$
2. $(y - 3)(y + 2)$
3. $(z - 5)(z - 1)$
4. $(m - 6)(3m + 4)$
5. $(5n + 3)(3n + 5)$
6. $(12w + 5)(2w - 1)$
7. $a(3a^2 + 10)$
8. $6b(8b^2 - 7)$
9. $5c(2c^2 + c - 9)$

### Example 2

Multiply $(3x - 7)(x^2 + 2x - 9)$.

## Solution

$$(3x - 7)(x^2 + 2x - 9) = 3x(x^2 + 2x - 9) - 7(x^2 + 2x - 9)$$
$$= 3x^3 + 6x^2 - 27x - 7x^2 - 14x + 63$$
$$= 3x^3 - x^2 - 41x + 63$$

**Multiply.**

10. $(a + 3)(a^2 + 2a + 4)$
11. $(b - 5)(3b^2 + 2b + 7)$
12. $(4c + 5)(c^2 - 3c + 4)$
13. $(2d - 1)(d^2 - 5d - 12)$
14. $(2m^2 - 5m + 3)(m + 2)$
15. $(6n^2 - 3n + 1)(4n - 1)$
16. $(3x^2 + 2x - 4)(5x + 9)$
17. $(2y^2 - 2y - 3)(3y - 2)$

### Example 3

Refer to the situation discussed in the Application.

a. Write expressions for the length and the width of the vegetable garden.

b. Write a polynomial expression for the area of the vegetable garden.

■ **Solution** ■

a. The length of the garden is made up of the sides of four squares and the widths of five walkways, so the total length of the garden is $(4x + 5y)$ ft. The width of the garden is made up of the side of one square and the widths of two walkways, so the total width of the garden is $(x + 2y)$ ft.

b. The area of a rectangle is found by multiplying length and width.

$A = (4x + 5y)(x + 2y)$
$\quad = 4x(x + 2y) + 5y(x + 2y)$
$\quad = 4x(x) + 4x(2y) + 5y(x) + 5y(2y)$ ← *Note*: These terms are in the order F, O, I, L.
$\quad = 4x^2 + 8xy + 5xy + 10y^2$
$\quad = 4x^2 + 13xy + 10y^2$

The area of the vegetable garden is $(4x^2 + 13xy + 10y^2)$ ft².

---

**Write a polynomial in standard form to represent the area of each figure described.**

**18.** a rectangle with length $(6x - 5)$ m and width $(2x + 7)$ m

**19.** a parallelogram with base $(4y + 9)$ ft and height $(3y + 2)$ ft

**20.** a triangle with base $(w + 10)$ cm and height $(5w + 2)$ cm

................
### Spiral Review

**For each system of equations, give the solution or state whether there are no solutions or many solutions.** *(Section 7.4)*

**21.** $2x - 3y = 3$
$\quad\ 5x + 2y = 17$

**22.** $1 = 3m - 7n$
$\quad\ -9 = 5m - n$

**23.** $a - 2b = 10$
$\quad\ -a + 2b = 12$

**Predict how the graph of each equation will compare with the graph of $y = x^2$. Explain how you made your decision.** *(Section 8.2)*

**24.** $y = 2x^2$

**25.** $y = -\frac{2}{5}x^2$

**26.** $y = -\frac{5}{4}x^2$

# Section 10.3 — Exploring Polynomial Equations

**GOAL**

**Learn how to . . .**
- factor out a linear factor from a polynomial

**So you can . . .**
- begin to use factoring to solve equations

## Application

A team of economists finds that the equation that models the relationship between the net sales $n$ of a certain computer and its retail price $p$ is

$$n = -100p^2 + 300{,}000p.$$

### Terms to Know | Example / Illustration

| Terms to Know | Example / Illustration |
|---|---|
| **Polynomial equation** (p. 436) — an equation in which both sides are polynomials | $5x^2 - 20x = 0$ |
| **Factoring a polynomial** (p. 437) — rewriting a polynomial as a product of polynomial factors | Factoring $5x^2 - 20x$ gives $5x(x - 4)$. |

## UNDERSTANDING THE MAIN IDEAS

You can solve a polynomial equation by getting all the terms on one side of the equals sign, factoring the resulting polynomial expression, and then using the following property.

*Zero-Product Property*

If a product of factors is equal to zero, then one or more of the factors is equal to zero.

The zero-product property is based on the fact that zero times any number is zero.

### Example 1

Factor each polynomial.

a. $30m^2 - 25m$

b. $49n^3 + 14n^2 - 7n$

### Solution

**a.** First find the greatest common factor (GCF) of the terms of the polynomial. Then rewrite each term as a product involving the common factor. Finally, factor out the GCF.

$30m^2 - 25m = 5m \cdot 6m - 5m \cdot 5$ ← The GCF of $30m^2$ and $25m$ is $5m$.

$\phantom{30m^2 - 25m} = 5m(6m - 5)$ ← Use the distributive property.

The factored form is $5m(6m - 5)$.

**b.** The GCF of $49n^3$, $14n^2$, and $7n$ is $7n$.

$49n^3 + 14n^2 - 7n = 7n \cdot 7n^2 + 7n \cdot 2n - 7n \cdot 1$

$\phantom{49n^3 + 14n^2 - 7n} = 7n(7n^2 + 2n - 1)$

**Factor each polynomial.**

1. $x^2 + 5x$
2. $3y^2 + 8y$
3. $2z^2 + 10z$
4. $12q^2 + 4q$
5. $20t^2 - 15t$
6. $v^3 - 3v^2 + v$

### Example 2

Solve the equation $2x(3x + 11) = 0$.

### Solution

$2x(3x + 11) = 0$ ← Use the zero-product property.

$2x = 0$ or $3x + 11 = 0$

$x = 0 \phantom{xxxx} 3x = -11$

$\phantom{x = 0 xxxx} x = -\dfrac{11}{3}$

The solutions are $-\dfrac{11}{3}$ and $0$.

**Solve each equation.**

7. $y(y - 1) = 0$
8. $3x(x + 4) = 0$
9. $0 = 5z(z - 7)$
10. $t(2t + 3) = 0$
11. $0 = 4n(5n - 1)$
12. $(m - 5)(m + 5) = 0$

Many polynomial equations must be factored before the zero-product property can be used.

### Example 3

Solve each equation.

a. $0 = 2x^2 + 3x$         b. $8y^2 = -2y$

**Solution**

a. $0 = 2x^2 + 3x$  ← Factor the right side; the GCF is $x$.
$0 = x(2x + 3)$  ← Use the zero-product property.
$x = 0$ or $2x + 3 = 0$
$\qquad\qquad 2x = -3$
$\qquad\qquad x = -\dfrac{3}{2}$

The solutions are $-\dfrac{3}{2}$ and 0.

b. $8y^2 = -2y$  ← Get all the terms on one side.
$8y^2 + 2y = 0$  ← Factor the left side; the GCF is $2y$.
$2y(4y + 1) = 0$  ← Use the zero-product property.
$2y = 0$ or $4y + 1 = 0$
$y = 0 \qquad\qquad 4y = -1$
$\qquad\qquad\qquad y = -\dfrac{1}{4}$

The solutions are $-\dfrac{1}{4}$ and 0.

**Solve each equation.**

13. $x^2 + 4x = 0$
14. $2y^2 + 5y = 0$
15. $3z^2 - 15z = 0$
16. $1 = 2n^2 - n$
17. $3m^2 + 4 = 8m$
18. $x^2 = 10x - 24$

........................
### Spiral Review

**Simplify each expression.** *(Section 9.4)*

19. $(15.66)^0$
20. $6^{-3}$
21. $\left(\dfrac{3}{4}\right)^{-4}$

**Multiply.** *(Section 10.2)*

22. $(2y + 1)(y + 7)$
23. $(3z - 4)(z - 4)$
24. $(3x - 7)(5x + 13)$

## Section 10.4: Exploring Factoring

**GOAL**

**Learn how to...**
- factor trinomials with positive quadratic terms and positive constant terms

**So you can...**
- use factoring to explore the value of polynomials

### Application

At the end of two years an investment of $500 at an annual growth rate of $r$, where $r$ is a decimal, will have a value given by the polynomial

$$500r^2 + 1000r + 500.$$

### UNDERSTANDING THE MAIN IDEAS

You have learned the FOIL method for multiplying two binomials. Factoring a trinomial whose terms have no common factors is the reverse of that process. If a trinomial is factorable, it will factor into the product of two binomials. To factor a trinomial, you use a process of guess, check, and revise. Try different combinations that will result in the correct quadratic and constant terms of the trinomial, focusing on finding the product that contains the correct linear term.

---

**Example 1**

Factor each trinomial.

   **a.** $x^2 + 4x + 3$           **b.** $x^2 - 7x + 10$

**Solution**

**a.** The first term is $x^2$, which has as its only factors $x$ and $x$. So the first term of each binomial is $x$.

$$x^2 + 5x + 6 = (x\quad)(x\quad)$$

Since the constant term 6 is *positive*, the two symbols in the binomials are the *same*. Further, since the linear term $5x$ has an addition symbol preceding it, the symbols must both be addition symbols.

$$x^2 + 5x + 6 = (x+\quad)(x+\quad)$$

*(Solution continues on next page.)*

### Solution (continued)

The factors of the constant term 6 are 2 and 3, and 1 and 6. We are looking for the pair whose *sum* is 5, the coefficient of the linear term. Since $2 + 3 = 5$, the factor pair 2 and 3 is the one we need.

$$x^2 + 5x + 6 = (x + 2)(x + 3)$$

You can check the factored form by using the FOIL method to multiply the two binomials: $x^2 + 3x + 2x + 6 = x^2 + 5x + 6$.

The correct factored form of $x^2 + 5x + 6$ is $(x + 2)(x + 3)$.

**b.** The first term is $x^2$, which has as its only factors $x$ and $x$. So the first term of each binomial is $x$.

$$x^2 - 7x + 10 = (x \quad )(x \quad )$$

Since the constant term 10 is *positive*, the two symbols in the binomials are the *same*. Since the linear term $7x$ has a subtraction symbol preceding it, the symbols must both be subtraction symbols.

$$x^2 - 7x + 10 = (x - \quad )(x - \quad )$$

The factors of the constant term 10 are 2 and 5, and 1 and 10. We are looking for the pair whose *sum* is 7, the coefficient of the linear term. Since $2 + 5 = 7$, the factor pair 2 and 5 is the one we need.

$$x^2 - 7x + 10 = (x - 2)(x - 5)$$

Check the factored form: $x^2 - 5x - 2x + 10 = x^2 - 7x + 10$.

The correct factored form of $x^2 - 7x + 10$ is $(x - 2)(x - 5)$.

---

**Factor, if possible. If not, write *not factorable*.**

1. $x^2 + 4x + 3$
2. $x^2 + 3x + 2$
3. $n^2 - 7n + 12$
4. $x^2 - 14x + 49$
5. $x^2 - 6x + 5$
6. $m^2 + 9m + 14$
7. $t^2 - 14t + 24$
8. $n^2 + 11n + 24$
9. $y^2 + 9y + 15$

### Example 2

Factor $2x^2 + 9x + 10$.

### Solution

The first term is $2x^2$, which has as its only factors $2x$ and $x$. So the first term of one binomial is $2x$ and the first term of the other is $x$.

$$2x^2 + 9x + 10 = (2x \quad )(x \quad )$$

*(Solution continues on next page.)*

> ### ■ Solution ■ (continued)
>
> Since the constant term 10 is *positive*, the two symbols in the binomials are the *same*. Since the linear term $9x$ has an addition symbol preceding it, the symbols must both be addition symbols.
>
> $$2x^2 + 9x + 10 = (2x + \phantom{0})(x + \phantom{0})$$
>
> The factors of the constant term 10 are 1 and 10, and 2 and 5. We are looking for the pair that yields a linear term of $9x$. Since the first terms of the binomials are not the same, there are two possible placements for each pair in the binomials above. Try each of the four possible binomial products to determine which arrangement gives the correct linear term.
>
> | | | |
> |---|---|---|
> | $(2x + 1)(x + 10)$ | $\rightarrow \quad 2x^2 + 20x + x + 10 = 2x^2 + 21x + 10$ | No |
> | $(2x + 10)(x + 1)$ | $\rightarrow \quad 2x^2 + 2x + 10x + 10 = 2x^2 + 12x + 10$ | No |
> | $(2x + 2)(x + 5)$ | $\rightarrow \quad 2x^2 + 10x + 2x + 10 = 2x^2 + 12x + 10$ | No |
> | $(2x + 5)(x + 2)$ | $\rightarrow \quad 2x^2 + 4x + 5x + 10 = 2x^2 + 9x + 10$ | Yes |
>
> So, the correct factored form of $2x^2 + 9x + 10$ is $(2x + 5)(x + 2)$.

**Factor each trinomial.**

10. $3x^2 + 5x + 2$
11. $5y^2 - 11y + 2$
12. $3z^2 + 11z + 10$
13. $4x^2 + 25x + 6$
14. $10n^2 - 27n + 5$
15. $15m^2 + 11m + 2$
16. $9t^2 + 18t + 8$
17. $6v^2 - 17v + 5$
18. $4x^2 + 11x + 6$

### Spiral Review

**Solve each equation algebraically.** *(Section 8.3)*

19. $2x^2 = 72$
20. $0 = y^2 - 81$
21. $-3z^2 + 47 = 20$

**Solve each equation using the quadratic formula.** *(Section 8.5)*

22. $x^2 + 8x + 12 = 0$
23. $3y^2 + 8y + 4 = 0$
24. $5z^2 - 13z + 8 = 0$

# Section 10.5 Applying Factoring

**GOAL**

**Learn how to . . .**
- factor quadratic polynomials with a negative constant term

**So you can . . .**
- solve quadratic equations by factoring

### Application

The height of a softball is modeled by the equation $h = -16t^2 + 24t + 2$, where $h$ is the height of the ball $t$ seconds after it is hit.

### UNDERSTANDING THE MAIN IDEAS

The pattern you use to factor a quadratic trinomial with a negative constant term is different than the pattern you use when the constant term is positive.

#### Example 1

Factor $5x^2 + 2x - 3$.

#### Solution

To factor $5x^2 + 2x - 3$, you should first notice that the constant term is negative.

The first term of the trinomial is $5x^2$, which has as its only factors $5x$ and $x$. So the first term of one binomial is $5x$ and the first term of the other is $x$.

$$5x^2 + 2x - 3 = (5x \quad )(x \quad )$$

Recall that the product of two numbers is negative if they have opposite signs. This means that the constant terms of the binomial factors must have different signs preceding them. Since the first terms of the binomial factors are different, we do not yet know which binomial will have a subtraction sign in it and which will have an addition sign.

$$5x^2 + 2x - 3 = (5x \pm \quad )(x \mp \quad )$$

*(Solution continues on next page.)*

Study Guide, ALGEBRA 1: EXPLORATIONS AND APPLICATIONS
Copyright © McDougal Littell Inc. All rights reserved.

### Solution (continued)

The only factors of the constant term 3 are 1 and 3. There are two possible placements of these factors and two choices for the signs in each arrangement, or a total of four possibilities to check.

$(5x + 1)(x - 3) \rightarrow 5x^2 - 15x + x - 3 = 5x^2 - 14x - 3$    No
$(5x - 1)(x + 3) \rightarrow 5x^2 + 15x - x - 3 = 5x^2 + 14x - 3$    No
$(5x + 3)(x - 1) \rightarrow 5x^2 - 5x + 3x - 3 = 5x^2 - 2x - 3$    No
$(5x - 3)(x + 1) \rightarrow 5x^2 + 5x - 3x - 3 = 5x^2 + 2x - 3$    Yes

So, the correct factored form of $5x^2 + 2x - 3$ is $(5x - 3)(x + 1)$.

**Replace each _?_ with the correct symbol or number.**

1. $x^2 + 6x - 7 = (x \underline{\ ?\ } 7)(x \underline{\ ?\ } 1)$
2. $2n^2 - 3n - 5 = (2n - \underline{\ ?\ })(n + \underline{\ ?\ })$
3. $6y^2 - 7y - 3 = (2y \underline{\ ?\ } 3)(3y \underline{\ ?\ } 1)$

**Factor each trinomial.**

4. $x^2 + 4x - 5$
5. $x^2 + x - 6$
6. $x^2 - 9x - 10$
7. $y^2 - 8y - 20$
8. $z^2 - z - 12$
9. $w^2 + 4w - 12$
10. $3x^2 - 14x - 5$
11. $2y^2 - 7y - 9$
12. $2z^2 + 5z - 7$

If an equation involves a polynomial set equal to zero (or can be rewritten in such a form) and the polynomial *can be factored*, you can use the zero-product property to solve the equation. (*Note*: You can tell that a quadratic equation is factorable if its discriminant is a perfect square.) If not, you can use the quadratic formula or graphing to solve the polynomial equation.

### Example 2

Solve $5x^2 + 2x = 3$ by factoring.

### Solution

Begin by writing the equation in standard form, $ax^2 + bx + c = 0$.

$5x^2 + 2x = 3$ ← Subtract 3 from both sides.

$5x^2 + 2x - 3 = 0$ ← Factor the left side. (See Example 1.)

$(5x - 3)(x + 1) = 0$ ← Use the zero-product property.

$5x - 3 = 0 \quad \text{or} \quad x + 1 = 0$

$5x = 3 \qquad\qquad x = -1$

$x = \dfrac{3}{5}$

The solutions are $-1$ and $\dfrac{3}{5}$.

**Solve each equation by factoring.**

13. $x^2 - 2x - 3 = 0$
14. $x^2 - 9x + 14 = 0$
15. $x^2 - 4x - 5 = 0$
16. $2x^2 - 5x - 18 = 0$
17. $2x^2 + 5x = 7$
18. $4x^2 - 25 = 0$

**Solve using graphing, the quadratic formula, or factoring.**

19. $2x^2 + 5x - 3 = 0$
20. $5x^2 + x = 0$
21. $x^2 - 3x = 40$
22. $2x^2 - x - 6 = 0$
23. $6x^2 = 24$
24. $5x^2 + 9x + 4 = 0$

### Example 3

The height of a softball is modeled by the equation $h = -16t^2 + 24t + 2$, where $h$ is the height of the ball $t$ seconds after it is hit. How long does it take the ball to reach a height of 10 ft?

### Solution

We want to know when $h = 10$.

$10 = -16t^2 + 24t + 2$

$0 = -16t^2 + 24t - 8$

$-\dfrac{1}{8}(0) = -\dfrac{1}{8}(-16t^2 + 24t - 8)$ ← Multiply both sides by $-\dfrac{1}{8}$.

$0 = 2t^2 - 3t + 1$ ← This equation is easier to factor.

$0 = (2t - 1)(t - 1)$

$2t - 1 = 0 \quad \text{or} \quad t - 1 = 0$

$2t = 1 \qquad\qquad t = 1$

$t = \dfrac{1}{2}$

*(Solution continues on next page.)*

> **Solution** (continued)
>
> The solutions of the equation are $t = \frac{1}{2}$ and $t = 1$. Each answer is a reasonable time for the situation, since the softball reaches 10 ft once going up to its maximum height and then again coming back down. Since the softball first reaches a height of 10 ft on the way up when $t = \frac{1}{2}$, the solution is $\frac{1}{2}$ s.

**25.** When is the softball 2 ft off the ground?

## Spiral Review

**Graph each point in a coordinate plane. Label each point with its letter. Name the quadrant (if any) in which the point lies.** *(Section 2.4)*

**26.** $A(2, -3)$  **27.** $B(-1, -2)$  **28.** $C(-2, 0)$  **29.** $D(-3, 1)$

**Tell whether the data show direct variation. Give reasons for your answer.** *(Section 3.2)*

**30.**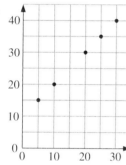

**31.**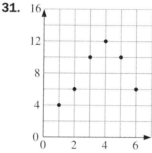

# Chapter 10 Review

**CHAPTER CHECK-UP**

Complete these exercises for a review of Chapter 10. If you have difficulty with a particular problem, review the indicated section.

**Add or subtract as indicated.** *(Section 10.1)*

1. $(2a^2 + 3a - 5) + (a^2 - 7a + 10)$
2. $(b^2 + 4b - 3) + (-3b - 2)$
3. $(8x^2 - 2x + 7) - (4x^2 + x - 1)$
4. $(3y^2 - 6y + 10) - (-5y^2 + 4)$

**Multiply.** *(Section 10.2)*

5. $(x + 4)(x + 6)$
6. $(y - 2)(y + 4)$
7. $(z - 3)(z - 1)$
8. $(a + 2)(a^2 + 2a + 3)$
9. $(b - 2)(3b^2 + 5b - 10)$
10. $(n + 1)(6n^2 - 5)$

**Factor each polynomial.** *(Section 10.3)*

11. $c^2 + 2c$
12. $4x^2 + 8x$
13. $45b^2 - 9b$

**Solve each equation.** *(Section 10.3)*

14. $3m(m + 7) = 0$
15. $0 = 2b(b - 9)$
16. $2n^2 + 5n = 0$
17. $2z^2 + 14z = 0$

**Factor, if possible. If not, write *not factorable*.** *(Sections 10.4 and 10.5)*

18. $x^2 + 9x + 14$
19. $x^2 + 2x + 2$
20. $n^2 - 7n + 10$
21. $3x^2 + 5x + 2$
22. $5y^2 - 11y + 2$
23. $3z^2 + 11z + 10$

**For Exercises 24–26, solve each equation by factoring.** *(Section 10.5)*

24. $p^2 - 2p - 3 = 0$
25. $4y^2 + 21y = 18$
26. $x^2 - 4x - 5 = 0$

27. An investment of $200 at an annual growth rate $r$, where $r$ is a decimal, will have a value at the end of the two years given by the polynomial $200r^2 + 400r + 200$. Write the polynomial in factored form and find the value at the end of the two years when $r = 0.05$. *(Section 10.5)*

**SPIRAL REVIEW    Chapters 1–10**

**Find each product.**

1. $(x - 7)(x + 7)$
2. $(3y - 5)(y + 3)$
3. $(5z - 2)(2z - 3)$

**Give the solution of each system or state whether there are *no solutions* or *many solutions*.**

4. $2x - 3y = -1$
   $5x + 2y = 7$
5. $27 = 3y - 7z$
   $13 = 5y - z$
6. $a - 2b = 10$
   $-a + 2b = 12$

**Estimate the solutions of each equation using the given graph.**

**7.** $x^2 - 5 = 0$

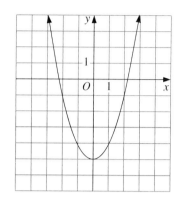

**8.** $-2x^2 + 32 = 0$

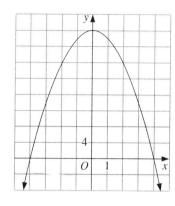

**9.** $\frac{1}{2}x^2 - 3 = 0$

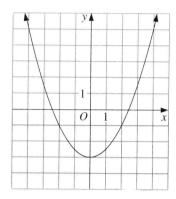

**Multiply.**

**10.** $(7y + 4)(y + 3)$

**11.** $(5z - 3)(z - 8)$

**12.** $(2y - 7)^2$

**Solve each equation.**

**13.** $2x^2 = 72$

**14.** $0 = y^2 - 81$

**15.** $-3z^2 + 47 = 20$

**16.** $x^2 + 8x + 12 = 0$

**17.** $3y^2 + 8y + 4 = 0$

**18.** $5z^2 - 13z + 8 = 0$

**Solve each equation for y and graph it.**

**19.** $3x - y = 9$

**20.** $2x + 3y = 6$

**21.** $3y - 2x = 9$

# Section 11.1 Exploring Inverse Variation

**GOAL**

**Learn how to...**
- recognize and describe data that show inverse variation

**So you can...**
- solve problems involving inverse variation

### Application
Three friends are planning a bicycle trip along a bike path from their school to a lake. They are deciding how much time they want to take for the trip and what average speed they should maintain. They can use inverse variation to relate their speed and time for the trip.

### Terms to Know | Example / Illustration

**Inverse variation (p. 464)**
a situation in which two variables have a constant, nonzero product

If $x$ and $y$ are variables, then the following equations model inverse variation.

$$x \cdot y = 7 \qquad y = \frac{7}{x} \qquad x = \frac{7}{y}$$

**Constant of variation (p. 464)**
the constant product in an inverse variation

The constant of variation for each of the inverse variation equations above is 7.

## UNDERSTANDING THE MAIN IDEAS

When the product of two variables is a nonzero constant, the two variables are in *inverse variation*. The nonzero constant is called the *constant of variation*.

Inverse variation can be described by an equation in the form $xy = k$ or $y = \frac{k}{x}$, where $k$ is the constant of variation ($k \neq 0$).

Recall that direct variation also has a constant of variation. Direct variation can be described by an equation in the form $\frac{y}{x} = k$ or $y = kx$, where $k$ is the constant of variation ($k \neq 0$).

### Example 1

Tell whether the data show *inverse variation*, *direct variation*, or *neither*. If the data show direct variation or inverse variation, write an equation for y in terms of x.

a.
| x | y |
|---|---|
| 3 | 15 |
| 4 | 20 |
| 5 | 25 |
| 6 | 30 |

b.
| x | y |
|---|---|
| 3 | 20 |
| 4 | 15 |
| 5 | 12 |
| 6 | 10 |

c.
| x | y |
|---|---|
| 3 | 14 |
| 4 | 17 |
| 5 | 20 |
| 6 | 23 |

### Solution

Check the data in the equations for inverse variation and direct variation.

**a.** inverse: $x \cdot y = k$      direct: $y = k \cdot x$

$3 \cdot 15 = k$, so $k = 45$     $15 = k \cdot 3$, so $k = 5$

$4 \cdot 20 = k$, so $k = 80$     $20 = k \cdot 4$, so $k = 5$

$5 \cdot 25 = k$, so $k = 125$     $25 = k \cdot 5$, so $k = 5$

$6 \cdot 30 = k$, so $k = 180$     $30 = k \cdot 6$, so $k = 5$

The data show direct variation. The constant of variation is $k = 5$. An equation for the data is $y = 5x$.

**b.** inverse: $x \cdot y = k$     direct: $y = k \cdot x$

$3 \cdot 20 = k$, so $k = 60$     $20 = k \cdot 3$, so $k = \frac{20}{3}$

$4 \cdot 15 = k$, so $k = 60$     $15 = k \cdot 4$, so $k = \frac{15}{4}$

$5 \cdot 12 = k$, so $k = 60$     $12 = k \cdot 5$, so $k = \frac{12}{5}$

$6 \cdot 10 = k$, so $k = 60$     $10 = k \cdot 6$, so $k = \frac{5}{3}$

The data show inverse variation. The constant of variation is $k = 60$. An equation for the data is $xy = 60$.

**c.** inverse: $x \cdot y = k$     direct: $y = k \cdot x$

$3 \cdot 14 = k$, so $k = 42$     $14 = k \cdot 3$, so $k = \frac{14}{3}$

$4 \cdot 17 = k$, so $k = 68$     $17 = k \cdot 4$, so $k = \frac{17}{4}$

$5 \cdot 20 = k$, so $k = 100$     $20 = k \cdot 5$, so $k = 4$

$6 \cdot 23 = k$, so $k = 138$     $23 = k \cdot 6$, so $k = \frac{23}{6}$

The data show neither inverse variation nor direct variation.

**Tell whether the data show *inverse variation*, *direct variation*, or *neither*.**
**If the data show inverse variation or direct variation, write an equation for $y$ in terms of $x$.**

1.
| $x$ | $y$ |
|---|---|
| 4 | 12 |
| 6 | 8 |
| 8 | 6 |
| 16 | 3 |

2.
| $x$ | $y$ |
|---|---|
| 4 | 5 |
| 6 | 9 |
| 8 | 13 |
| 16 | 29 |

3.
| $x$ | $y$ |
|---|---|
| 10 | 10 |
| 20 | 5 |
| 25 | 4 |
| 50 | 2 |

4.
| $x$ | $y$ |
|---|---|
| 10 | 6 |
| 20 | 12 |
| 25 | 15 |
| 40 | 24 |

### Example 2

These data show inverse variation. Find the constant of variation. Write equations relating $y$ and $x$ in the form $xy = k$ and in the form $y = \dfrac{k}{x}$.

a.
| $x$ | $y$ |
|---|---|
| $-3$ | 4 |
| $-1$ | 12 |
| 1 | $-12$ |
| 6 | $-2$ |

b.
| $x$ | $y$ |
|---|---|
| $-2$ | $-1$ |
| $-1$ | $-2$ |
| 4 | $\dfrac{1}{2}$ |
| 6 | $\dfrac{1}{3}$ |

**Solution**

a. In each row, the product $xy$ is $-12$. So the constant of variation is $-12$.
The equations are $xy = -12$ and $y = -\dfrac{12}{x}$.

b. In each row, the product $xy$ is 2. So the constant of variation is 2.
The equations are $xy = 2$ and $y = \dfrac{2}{x}$.

**The data show inverse variation. Find the missing values.**

5.
| $x$ | $y$ |
|---|---|
| $-12$ | ? |
| $-4$ | 9 |
| ? | 36 |
| 3 | ? |

6.
| $x$ | $y$ |
|---|---|
| 1 | ? |
| 2 | ? |
| 8 | $\dfrac{1}{2}$ |
| ? | $\dfrac{2}{5}$ |

**Write each inverse variation equation in the form $y = \dfrac{k}{x}$.**

**7.** $xy = 15$      **8.** $xy = -3$      **9.** $xy = 1$      **10.** $xy = c$

**Use the inverse variation equation $y = \dfrac{k}{x}$ to find each value.**

**11.** Find $y$ if $k = -20$ and $x = -10$.

**12.** Find $x$ if $y = 3$ and $k = 30$.

**13.** Find $k$ if $y = -\dfrac{11}{2}$ and $x = -\dfrac{2}{3}$.

## *Graphing inverse variation equations*

The graph of an inverse variation equation always has two parts (sometimes called branches) located in "opposite" quadrants. These branches approach the axes, but never intersect them.

### Example 3

Graph the inverse variation equation $xy = -10$.

**■ Solution ■**

Make a table of values for $x$ and $y$. Rewriting $xy = -10$ as $y = \dfrac{-10}{x}$ may be helpful.

| $x$ | $y$ |
|---|---|
| $-10$ | $1$ |
| $-5$ | $2$ |
| $-4$ | $2\frac{1}{2}$ |
| $-3$ | $3\frac{1}{3}$ |
| $-2$ | $5$ |
| $-1$ | $10$ |

| $x$ | $y$ |
|---|---|
| $1$ | $-10$ |
| $2$ | $-5$ |
| $2\frac{1}{2}$ | $-4$ |
| $3\frac{1}{3}$ | $-3$ |
| $5$ | $-2$ |
| $10$ | $-1$ |

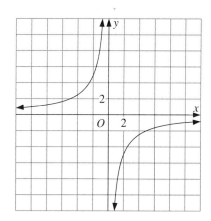

The graph is shown at the right.

**For Exercises 14–17, match each inverse variation equation with one of the two graphs.**

14. $xy = 5$

15. $y = \dfrac{-5}{x}$

16. $xy = -5$

17. $y = \dfrac{5}{x}$

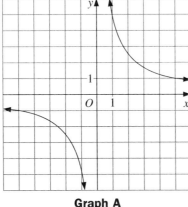

Graph A

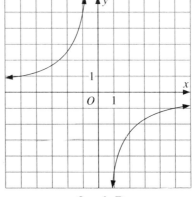

Graph B

18. **Writing** Think about a trip you have taken many times, such as a walk from your house to a place in your town. Explain how your speed on the trip is related to the time needed to make the trip.

## Spiral Review

**Find the mean, the median, and the mode(s) of each set of data.** *(Section 1.3)*

19. 80, 80, 90, 80, 90

20. 10.3, 10.4, 10.2, 10.2, 10.7, 10.0

**Solve each system of equations by adding or subtracting.** *(Section 7.3)*

21. $5a + \dfrac{1}{2}b = 28$
    $3a + \dfrac{1}{2}b = 18$

22. $4p + 3s = -7$
    $-4p + s = -13$

# Section 11.2 Using Weighted Averages

**GOAL**

**Learn how to...**
- find weighted averages

**So you can...**
- summarize information such as grades or scores

## Application

The grading policy of the mathematics department at a school states that a chapter test score counts 3 times as much as a quiz score and that the score on the semester test counts 2 times as much as a chapter test score. The mathematics teachers use weighted averages to assign semester grades based on each student's scores on their quizzes, chapter tests, and semester test.

### Terms to Know

**Weighted average (p. 470)**
the mean of data when some of the data values count more than others.

### Example / Illustration

| Score | Weight |
|-------|--------|
| 8     | 1      |
| 9     | 2      |
| 13    | 1      |
| 15    | 3      |

The weighted average is

$$\frac{8 \cdot 1 + 9 \cdot 2 + 13 \cdot 1 + 15 \cdot 3}{7} = \frac{84}{7}$$
$$= 12$$

Notice that you divide by the *sum of the weights*, not the number of scores.

## UNDERSTANDING THE MAIN IDEAS

### Using repeated numbers in data

Sometimes when you find an average, you may want to give more weight to one number (or several numbers) than the others.

### Example 1

Refer to the Application. Suppose you have quiz scores of 85, 85, 90, 95, and 100, chapter test scores of 82, 85, and 92 and a semester test score of 88. Based on the weighting system given, what is your weighted average?

### Solution

The weight for each quiz is 1, so the weight of each chapter test is $3 \times 1 = 3$ and the weight for the semester test is $2 \times 3 = 6$.

The numerator in the weighted average is the sum of the products of each score and its weight. The denominator in the weighted average is the sum of the weights.

$$\frac{85 \cdot 1 + 85 \cdot 1 + 90 \cdot 1 + 95 \cdot 1 + 100 \cdot 1 + 82 \cdot 3 + 85 \cdot 3 + 92 \cdot 3 + 88 \cdot 6}{1 + 1 + 1 + 1 + 1 + 3 + 3 + 3 + 6}$$

$$= \frac{85 + 85 + 90 + 95 + 100 + 246 + 255 + 276 + 528}{20}$$

$$= \frac{1760}{20}$$

$$= 88$$

Your weighted average is 88.

**For Exercises 1 and 2, find each weighted average.**

1.  
| Score | Weight |
|---|---|
| 15 | 1 |
| 17 | 3 |
| 18 | 4 |
| 19 | 3 |
| 33 | 1 |

2.  
| Score | Weight |
|---|---|
| 20 | 3 |
| 22 | 5 |
| 23 | 4 |
| 28 | 2 |
| 29 | 2 |

3. Refer to the weighting system given in the Application. Suppose a student has quiz scores of 78, 80, 81, 84, and 92, chapter test scores of 84, 88, and 93, and a semester test score of 92. Find the student's weighted average.

### Example 2

A student's test scores are 87, 85, and 92. There is one more test and it counts twice as much as the others. What score does the student need so that the weighted test average is 90?

> ### ■ Solution ■
>
> Let $x$ = the score on the remaining test. Since this score counts twice as much as the others, multiply it by 2.
>
> $\dfrac{87 + 85 + 92 + 2 \cdot x}{5} = 90$ ← The weighted average is to be 90. The sum of the weights is $1 + 1 + 1 + 2 = 5$.
>
> $\dfrac{264 + 2x}{5} = 90$ ← Multiply both sides of the equation by 5.
>
> $264 + 2x = 450$ ← Subtract 264 from both sides.
>
> $2x = 186$ ← Divide both sides by 2.
>
> $x = 93$
>
> The student needs a score of 93 on the remaining test to have an average of 90.

**A student's scores on five tests are 83, 88, 82, 91, and 90. There is one test remaining.**

4. Suppose the sixth test counts twice as much as the others. Can the student get an average of 90 for all the tests? If so, what score does the student need on the sixth test?

5. Suppose the sixth test counts three times as much as the others. Can the student get an average of 93 for all the tests? If so, what score does the student need on the sixth test?

6. **Mathematics Journal** Make a list of the movies or books you have liked and disliked during the past year. Think about how you could use a weighted average to find out which movie or book represents the "average" of your like-and-dislike list.

......................
### Spiral Review

**Solve each equation.** *(Section 4.4)*

7. $\dfrac{x}{5} - \dfrac{2}{3}x = 7$

8. $y + \dfrac{y}{2} - \dfrac{y}{3} = 1$

9. $\dfrac{n}{4} - \dfrac{n}{5} = -\dfrac{7}{10}$

**Tell whether each expression is a polynomial. Write *yes* or *no*. If not, explain why not.** *(Section 10.1)*

10. $5x + 3 - 2x^2$

11. $\dfrac{4}{5}x^2 y^3$

12. $3\left(\dfrac{1}{x}\right)^2 + 5\left(\dfrac{1}{x}\right) - 2$

# Section 11.3 Solving Rational Equations

**GOAL**

**Learn how to . . .**
- solve rational equations

**So you can . . .**
- solve problems involving rational expressions

### Application

Did you know that you weigh less as you climb a mountain or ascend in an airplane? The formula

$$w = \frac{(4000)^2 s}{(4000 + h)^2}$$

gives the weight $w$ of a person who is $h$ mi above sea level and whose weight at sea level is $s$ lb.

Many formulas involve rational expressions.

### Terms to Know / Example / Illustration

| Terms to Know | Example / Illustration |
|---|---|
| **Rational expression (p. 476)** an expression that can be written as a fraction (The numerator and denominator must be polynomials.) | $\dfrac{5}{2x} \qquad \dfrac{x+3}{x+2}$ |
| **Rational equation (p. 476)** an equation that contains only rational expressions | $\dfrac{1}{3x+5} = \dfrac{4x^2 + 2x + 1}{12x - 3}$ |

## UNDERSTANDING THE MAIN IDEAS

The same skills you use to find the least common denominator (LCD) for fractions are used to find the LCD for rational expressions.

### Example 1

Find the least common denominator for each group of rational expressions.

**a.** $\dfrac{3}{x}, \dfrac{5}{x+1},$ and $\dfrac{12}{9x}$   **b.** $\dfrac{6}{25x}, \dfrac{x+1}{x^2},$ and $\dfrac{8x}{5(x-2)}$

■ **Solution** ■

**a.** The denominator $9x$ can be rewritten as $3 \cdot 3 \cdot x$.
The LCD is $3 \cdot 3 \cdot x \cdot (x + 1)$, or $9x(x + 1)$.

**b.** The denominator $25x$ can be rewritten as $5 \cdot 5 \cdot x$ and $x^2$ can be rewritten as $x \cdot x$.
The LCD is $5 \cdot 5 \cdot x \cdot x \cdot (x - 2)$, or $25x^2(x - 2)$.

**Find the least common denominator for each group of rational expressions.**

1. $\dfrac{2x}{3}, \dfrac{3x}{2}, \dfrac{5}{3x}$

2. $\dfrac{3}{5(x+5)}, \dfrac{5x}{2}, \dfrac{9x}{10}, \dfrac{3x+4}{x+5}$

3. $\dfrac{7}{3m+1}, \dfrac{m}{2}$

4. $\dfrac{7}{t^2}, \dfrac{8t+3}{t+1}, \dfrac{5-3t}{t}$

## Solving rational equations

One way to solve a rational equation is to begin by multiplying both sides of the equation by the LCD of all the rational expressions. This results in an equation that does not contain any fractions.

### Example 2

Solve the equation $\dfrac{3}{x} + \dfrac{5}{x+1} = \dfrac{12}{9x}$.

**Solution**

From Example 1, the LCD is $9x(x+1)$.

Multiply both sides of the equation by $9x(x+1)$.

$$\dfrac{3}{x} + \dfrac{5}{x+1} = \dfrac{12}{9x}$$

$$9x(x+1)\left(\dfrac{3}{x} + \dfrac{5}{x+1}\right) = 9x(x+1)\left(\dfrac{12}{9x}\right)$$

$$\dfrac{9x(x+1) \cdot 3}{x} + \dfrac{9x(x+1) \cdot 5}{x+1} = \dfrac{9x(x+1) \cdot 12}{9x} \quad \leftarrow \text{Cancel like factors in the numerator and denominator.}$$

$$27(x+1) + (9x)5 = 12(x+1)$$

$$27x + 27 + 45x = 12x + 12$$

$$72x + 27 = 12x + 12$$

$$60x = -15$$

$$x = -\dfrac{1}{4}$$

**Consider the rational equation** $\dfrac{5}{3a+1} - \dfrac{2}{a} = \dfrac{3}{3a+1}$.

5. What is the LCD of the fractions in the equation?

6. Multiply both sides of the equation by your answer to Exercise 5, and simplify both sides of the resulting equation.

7. What value of $a$ is the solution of the rational equation?

Consider the rational equation $\dfrac{m+3}{5m} = \dfrac{3}{5m} + \dfrac{20}{m^2}$.

8. What is the LCD of the fractions in the equation?

9. Multiply both sides of the equation by your answer to Exercise 8, and simplify both sides of the resulting equation.

10. What value of $m$ is the solution of the rational equation?

Solve each equation. If the equation has no solution, write *no solution*.

11. $\dfrac{6}{n} + 3 = \dfrac{24}{n}$

12. $\dfrac{x-1}{8} + \dfrac{x+1}{12} = 1$

13. $\dfrac{27}{a+5} + 4 = \dfrac{63}{a+5}$

14. $\dfrac{x-4}{8} = \dfrac{6}{x+4}$

Refer to the formula $w = \dfrac{(4000)^2 s}{(4000+h)^2}$ in the Application.

15. If a person weighs 165 lb at sea level, what is the person's apparent weight at 3 mi above sea level?

16. What is that person's apparent weight 10 mi above sea level?

......................
## Spiral Review

**Graph each equation.** *(Section 3.4)*

17. $y = \dfrac{2}{3}x + 4$

18. $y = -3x$

19. $y = 2x - 3$

**Solve each equation by factoring.** *(Section 10.5)*

20. $x^2 + 5x - 14 = 0$

21. $2a^2 - a - 1 = 0$

22. $4n^2 = 11n - 6$

# Section 11.4 Applying Formulas

**GOAL**

**Learn how to...**
- solve a formula for one of its variables

**So you can...**
- perform multiple calculations with the same formula
- make calculations with spreadsheets

## Application

On the Fahrenheit temperature scale, the freezing point of water is 32°F and the boiling point of water is 212°F. On the Celsius scale, water freezes at 0°C and boils at 100°C.

If you are given a Celsius temperature, you can convert it to a Fahrenheit temperature by using the formula

$$F = \frac{9}{5}C + 32$$

which gives the Fahrenheit temperature, $F$, in terms of the Celsius temperature, $C$. If you want to convert Fahrenheit temperatures to Celsius temperatures, you can solve the formula for $C$ in terms of $F$.

## UNDERSTANDING THE MAIN IDEAS

### Using formulas

A formula is an equation in which one variable is expressed in terms of constants, operation symbols, and other variables. If you know a value for all but one of the variables in a formula, you can calculate the value of the remaining variable.

### Example

Use the conversion formula $F = \frac{9}{5}C + 32$.

a. "Room temperature" is about 72°F. Is 15°C warmer or colder than this "room temperature?"

b. What is the equivalent Celsius temperature for a hot summer's day temperature of 95°F?

c. Solve the formula $F = \frac{9}{5}C + 32$ for $C$, and confirm your answer to part (b).

### ■ Solution ■

**a.** Use the given formula to convert 15°C to its Fahrenheit equivalent, and compare the result to 72°F.

$$F = \frac{9}{5}C + 32$$

$$= \frac{9}{5}(15) + 32 \quad \leftarrow \text{Substitute 15 for } C.$$

$$= 27 + 32$$

$$= 59$$

Since 15°C is equivalent to 59°F, a temperature of 15°C is colder than "room temperature."

**b.** Substitute 95 for $F$ in the formula and solve for $C$.

$$F = \frac{9}{5}C + 32$$

$$95 = \frac{9}{5}C + 32 \quad \leftarrow \text{Substitute 95 for } F.$$

$$63 = \frac{9}{5}C \quad \leftarrow \text{Subtract 32 from both sides of the equation.}$$

$$\frac{5}{9}(63) = \frac{5}{9}\left(\frac{9}{5}C\right) \quad \leftarrow \text{Multiply both sides by } \frac{5}{9}.$$

$$35 = C$$

The Celsius equivalent of 95°F is 35°C.

**c.**
$$F = \frac{9}{5}C + 32$$

$$F - 32 = \frac{9}{5}C \quad \leftarrow \text{Subtract 32 from both sides of the equation.}$$

$$\frac{5}{9}(F - 32) = \frac{5}{9}\left(\frac{9}{5}C\right) \quad \leftarrow \text{Multiply both sides of the equation by } \frac{5}{9}.$$

$$\frac{5}{9}(F - 32) = C \quad \text{or} \quad C = \frac{5}{9}(F - 32)$$

Now, substitute 95 for $F$ to verifty the answer in part (b).

$$C = \frac{5}{9}(95 - 32)$$

$$= \frac{5}{9}(63)$$

$$= 35$$

**Study Guide,** ALGEBRA 1: EXPLORATIONS AND APPLICATIONS

**Use the given formula to find the value of the variable.**

1. Use the area formula $A = \pi r^2$ to find the area $A$ of a circle when the radius $r$ is 10 in. Use $\pi \approx 3.14$.
2. Use the distance formula $d = r \cdot t$ to find the distance $d$ that you travel if you drive at an average rate $r$ of 55 mi/h for a time $t$ of 6 h.
3. Use the distance formula $d = r \cdot t$ to find the rate $r$ if you travel $d = 50$ km in $t = 17$ min.
4. Use the distance formula to find $r$ (in mi/h) if $d = 30$ mi and $t = 30$ min.

**For Exercises 5–14, use the formulas for converting between degrees Celsius and degrees Fahrenheit to find each temperature.**

5. What is the equivalent Fahrenheit temperature for 0°C?
6. What is the equivalent Fahrenheit temperature for 20°C?
7. What is the equivalent Fahrenheit temperature for 40°C?
8. What is the equivalent Fahrenheit temperature for –50°C?
9. What is the equivalent Fahrenheit temperature for –40°C?
10. What is the equivalent Celsius temperature for –4°F?
11. What is the equivalent Celsius temperature for 59°F?
12. What is the equivalent Celsius temperature for 98.6°F?
13. What is the equivalent Celsius temperature for 212°F?
14. What is the equivalent Celsius temperature for –40°F?

15. **Mathematics Journal** Thinking about your hobbies or favorite sports, what formulas are used in those activities? Describe the formulas and tell how they are used.

## Spiral Review

**Find the reciprocal of each number.** *(Section 4.2)*

16. $-\dfrac{12}{7}$
17. 10
18. $\dfrac{21}{22}$

**Simplify each expression.** *(Section 9.7)*

19. $(2m^2)^4$
20. $\left(\dfrac{a^2}{b^3}\right)^5$
21. $\dfrac{a^2 b^3 c}{a^3 b c^2}$

# Section 11.5 — Working with Rational Expressions

**GOAL**

**Learn how to . . .**
- multiply and divide rational expressions

**So you can . . .**
- solve problems involving rational expressions

*Application*

How much of the target is covered by the bull's-eye? The area of the bull's-eye is $\pi r^2$ and the area of the whole target is $\pi R^2$. You can use a rational expression in its simplest form to compare the area of the bull's-eye to the area of the entire target.

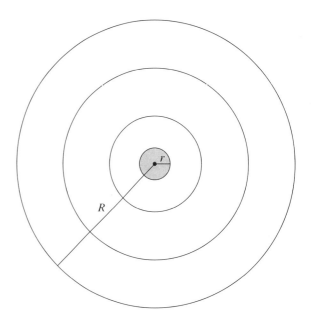

**Terms to Know**

**Simplest form of a rational expression (p. 489)**
a rational expression for which the greatest common factor of the numerator and denominator is 1

**Example / Illustration**

The rational expression $\dfrac{5x + y}{5}$ is in simplest form, while $\dfrac{5a^2b}{10ab}$ is not because the numerator and denominator have $5ab$ as a common factor.

## UNDERSTANDING THE MAIN IDEAS

### Multiplying rational expressions

You multiply rational expressions the same way you multiply fractions. That is, you multiply the numerators and you multiply the denominators separately. Before you multiply however, you can divide out the common factors of the numerators and denominators.

## Example 1

Find each product.

**a.** $\dfrac{6m^3}{5a} \cdot \dfrac{25a^4}{2m}$

**b.** $\dfrac{20x^3}{5x-20} \cdot \dfrac{3x-12}{6x^4}$

### Solution

**a.** Write the product with one numerator and one denominator. You can rearrange the factors in the numerator and denominator to make it easier to divide out common factors.

$\dfrac{6m^3}{5a} \cdot \dfrac{25a^4}{2m} = \dfrac{6m^3 \cdot 25a^4}{5a \cdot 2m}$ ← Write the product as one fraction.

$= \dfrac{2 \cdot 3 \cdot m \cdot m^2 \cdot 5 \cdot 5 \cdot a \cdot a^3}{5 \cdot a \cdot 2 \cdot m}$ ← Look for common factors.

$= \dfrac{2 \cdot 5 \cdot a \cdot m \cdot 3 \cdot m^2 \cdot 5 \cdot a^3}{2 \cdot 5 \cdot a \cdot m}$ ← You can rearrange the factors.

$= \dfrac{3 \cdot m^2 \cdot 5 \cdot a^3}{1}$ ← Divide out the common factors.

$= 15a^3m^2$

**b.** $\dfrac{20x^3}{5x-20} \cdot \dfrac{3x-12}{6x^4} = \dfrac{20x^3 \cdot (3x-12)}{(5x-20) \cdot 6x^4}$

$= \dfrac{20x^3 \cdot 3(x-4)}{5(x-4) \cdot 6x^4}$

$= \dfrac{5 \cdot 2 \cdot 2 \cdot x^3 \cdot 3(x-4)}{5(x-4) \cdot 2 \cdot 3 \cdot x^3 \cdot x}$

$= \dfrac{5 \cdot 2 \cdot 3 \cdot x^3 \cdot (x-4) \cdot 2}{5 \cdot 2 \cdot 3 \cdot x^3 \cdot (x-4) \cdot x}$

$= \dfrac{2}{x}$

**Factor each expression.**

**1.** $15a + 15b + 15c$

**2.** $12x - 18$

**3.** $15x^3 - 25x^2 + 5x$

**4.** $7(x+3)^2 + 5(x+3)$

**Simplify each expression.**

**5.** $\dfrac{3a}{2b} \cdot \dfrac{10b^5}{12a^2}$

**6.** $\dfrac{3x+9}{y+1} \cdot \dfrac{10y+10}{5x+15}$

**7.** $\dfrac{x^2-y^2}{3(x-y)} \cdot \dfrac{1}{x+y}$

**8.** $\dfrac{x^2-1}{6x^2} \cdot \dfrac{3x^2-3x}{(x-1)^2}$

## Dividing rational expressions

To divide one rational expression by a second rational expression, you should multiply by the reciprocal of the second rational expression. This is the same procedure that you use to divide fractions.

### Example 2

Simplify the expression $\dfrac{(m+4)^3}{-3m-12} \div \dfrac{m^2-16}{12}$.

### Solution

Begin by using the reciprocal of the second rational expression to rewrite the division as a multiplication.

$$\dfrac{(m+4)^3}{-3m-12} \div \dfrac{m^2-16}{12} = \dfrac{(m+4)^3}{-3m-12} \cdot \dfrac{12}{m^2-16}$$

$$= \dfrac{(m+4)^3(12)}{(-3m-12)(m^2-16)}$$

$$= \dfrac{(m+4)(m+4)(m+4)(3)(4)}{-3(m+4)(m+4)(m-4)}$$

$$= \dfrac{\cancel{3}\cancel{(m+4)}\cancel{(m+4)}(4)(m+4)}{\cancel{3}\cancel{(m+4)}\cancel{(m+4)}(-1)(m-4)}$$

$$= \dfrac{4(m+4)}{-1(m-4)}$$

$$= \dfrac{4m+16}{-m+4}$$

**Simplify each rational expression. If it is already in simplest form, write *simplest form*.**

9. $\dfrac{10}{a} \div \dfrac{2}{a^4}$

10. $(a+1)^2 \div (5a+5)$

11. $\dfrac{5a^2+5a+1}{5a}$

12. $\dfrac{st+5t}{t^2} \div \dfrac{7s+35}{14t^3}$

13. $\dfrac{5x-5y}{x^2-y^2} \div \dfrac{25}{10x+10y}$

14. $\dfrac{ab^2-ba^2}{2b-2a} \div \dfrac{ab(b-a)}{2b-2a}$

**Refer to the Application.**

15. If the radius of the bull's-eye is 2 in. and the radius of the entire target is 18 in., what fraction of the target is covered by the bull's-eye?

16. How does your answer to Exercise 15 compare to $\dfrac{2}{18} = \dfrac{1}{9}$, the ratio of the radii of the two circles?

**17. Open-ended Problem** Find a target with a bull's-eye, such as an archery or darts target. What fraction of the target is covered by the bull's-eye?

## Spiral Review

**Use the distributive property to find each product.** *(Section 6.4)*

**18.** $5x(x-3)$  **19.** $(2y+3)(2y-5)$  **20.** $(5+2\sqrt{3})(5-2\sqrt{3})$

**Simplify each expression.** *(Section 9.6)*

**21.** $m^2 \cdot m^5 \cdot m$  **22.** $\dfrac{t^5}{t^3}$  **23.** $\dfrac{p^{-5}}{p^{-11}}$

# Section 11.6 Exploring Rational Expressions

**GOAL**

**Learn how to...**
- add and subtract rational expressions

**So you can...**
- explore patterns with fractions
- solve problems

## Application

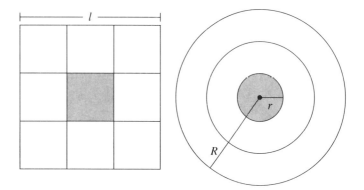

On which target do you have a better chance to get a bull's-eye? In the previous section, you saw how to use rational expressions to compare the area of a bull's-eye to the area of an entire target. You can subtract two rational expressions to compare them.

## UNDERSTANDING THE MAIN IDEAS

### Adding and subtracting rational expressions

Adding and subtracting rational expressions is similar to adding and subtracting fractions. When the denominators of the expressions are different, you rewrite each expression using their least common denominator (LCD) before adding or subtracting.

---

**Example 1**

Simplify each expression.

a. $\dfrac{a+5}{a+2} + \dfrac{a-1}{a+2}$

b. $\dfrac{3b}{b+6} - \dfrac{b-6}{2b}$

---

### Solution

**a.** The two rational expressions already have a common denominator.

$$\frac{a+5}{a+2} + \frac{a-1}{a+2} = \frac{a+5+a-1}{a+2}$$

$$= \frac{2a+4}{a+2} \quad \leftarrow \text{Combine like terms.}$$

$$= \frac{2(a+2)}{a+2} \quad \leftarrow \text{There is a common factor 2 in the numerator.}$$

$$= 2$$

**b.** The LCD is $2b(b+6)$.

$$\frac{3b}{b+6} - \frac{b-6}{2b} = \frac{3b}{b+6} \cdot \frac{2b}{2b} - \frac{b-6}{2b} \cdot \frac{b+6}{b+6}$$

$$= \frac{6b^2}{2b^2 + 12b} - \frac{b^2 - 36}{2b^2 + 12b}$$

$$= \frac{6b^2 - (b^2 - 36)}{2b^2 + 12b}$$

$$= \frac{6b^2 - b^2 + 36}{2b^2 + 12b}$$

$$= \frac{5b^2 + 36}{2b^2 + 12b}$$

**Simplify each expression.**

1. $\dfrac{5m}{m+3} + \dfrac{15}{m+3}$

2. $\dfrac{4x+7}{2x} - \dfrac{2x-3}{x}$

3. $\dfrac{1}{n-1} - \dfrac{1}{n}$

4. $\dfrac{10}{ab^2} + \dfrac{10}{a^2b}$

5. $\dfrac{2x+5y}{2x+3} - \dfrac{5y+2x}{3+2x}$

6. $\dfrac{2p}{p+5} + \dfrac{6}{p-2}$

## Types of functions

You have studied five types of functions in this book. Remember that for any function, each value of $x$ gives you a single value of $y$.

### Example 2

Describe each function by telling what type of function it is and how variables and/or exponents are used in the function.

**a.** $y = 3x^2 - 2x + 5$  **b.** $y = 5(2.0)^x$  **c.** $y = -\frac{3}{4}x + 2$

**d.** $y = \frac{10}{x+5} - 15$  **e.** $y = x^4 + 2x^3 - 5x^2 + 5x - 7$

### ■ Solution ■

**a.** The greatest exponent is 2. It is a *quadratic function*.

**b.** A variable appears as an exponent. It is an *exponential function*.

**c.** The greatest exponent is 1. It is a *linear function*.

**d.** A variable appears in a denominator. It is a *rational function*.

**e.** The exponents are whole numbers, one of which is greater than 3. It is a *polynomial function*.

---

**State whether each function can be best described as *linear*, *quadratic*, *polynomial*, *rational*, or *exponential*.**

**7.** $y = 3x^2 - 5x + 7$    **8.** $y = \frac{3}{x}$    **9.** $y = 3.5x - 2.5$

**10.** $y = 3^x$    **11.** $y = \frac{15}{x} + 7$    **12.** $y = 3x^3 + x^2 - 3$

**13.** $y = 3x$    **14.** $y = 5(0.2)^x$    **15.** $y = 3x^2$

**Refer to the Application.**

**16.** If $l = 6$, $r = 1$, and $R = 4$, find the fraction of each target that is covered by its bull's-eye.

**17.** On which target do you have a better chance to hit a bull's-eye? Explain your answer.

### Spiral Review

**Solve each equation. If the equation has no solution, write *no solution*.** *(Section 11.3)*

**18.** $2 - \frac{5}{x} = \frac{4}{x} + 3$    **19.** $\frac{8}{x+1} = \frac{12}{x+6}$

**Solve each equation for x.** *(Section 11.4)*

**20.** $A = \frac{1}{2}b(x+y)$    **21.** $\frac{5}{x} - \frac{4}{y} = \frac{5}{z}$

# Chapter 11 Review

**CHAPTER CHECK-UP**

Complete these exercises for a review of Chapter 11. If you have difficulty with a particular problem, review the indicated section.

For Exercises 1–3, tell whether the data show *inverse variation*, *direct variation*, or *neither*. *(Section 11.1)*

1.
| x | y |
|---|---|
| −1 | 5 |
| 2 | −1 |
| 4 | 5 |

2.
| x | y |
|---|---|
| −1 | −12 |
| 2 | 6 |
| 4 | 3 |

3.
| x | y |
|---|---|
| −1 | 2 |
| 2 | −4 |
| 4 | −8 |

4. The data below show inverse variation. Find the constant of variation $k$ and write an equation in the form $xy = k$. *(Section 11.1)*

| x | y |
|---|---|
| −4 | −9 |
| −2 | −18 |
| 3 | 12 |

5. Find the weighted average of these scores. *(Section 11.2)*

| Score | Weight |
|---|---|
| 8 | 6 |
| 9 | 10 |
| 11 | 10 |
| 13 | 4 |

6. Find the least common denominator for the expressions $\frac{7x+1}{5x}$, $\frac{1}{2}$, and $\frac{2x+3}{2x}$. *(Section 11.3)*

7. Solve the equation $\frac{7x+1}{5x} - \frac{1}{2} = \frac{2x+3}{2x}$. *(Section 11.3)*

8. Using the formula $F = \frac{9}{5}C + 32$, find the value of $C$ when $F = -4$. *(Section 11.4)*

**Simplify each expression.** *(Sections 11.5 and 11.6)*

9. $\frac{5x-3}{2x} \cdot \frac{x^2}{10x-6}$

10. $\frac{x^2-4}{x-2} \div \frac{x+2}{5}$

11. $\frac{3}{2m} + \frac{5}{3m}$

12. $\frac{1}{n+2} - \frac{1}{5}$

**SPIRAL REVIEW**  Chapters 1–11

**Simplify each expression.**

1. $3 + 2 \times 5 - 4 \times 3$
2. $|5 - 3| - |-7| + |4(-3)|$
3. $(2n + 5)^2$
4. $(x - z)(x + z)$
5. $(3 + \sqrt{5})(5 + \sqrt{5})$
6. $\dfrac{(3a)^3}{(5a)^0}$
7. $3(2p^2 + 4p + 2) - 5(3p^2 - 9p + 1)$

**For Exercises 8–15, solve each equation.**

8. $-12x = 180$
9. $5y - 7 = -12$
10. $\dfrac{2}{5}y = 7$
11. $4(x + 2) + 3x = 7(5 - x) - (4x - 1)$
12. $8.6x - 7.1 = -2.8$
13. $x^2 - 5x + 4 = 0$
14. $x^2 + 3x - 12 = 0$
15. $\dfrac{m}{5} = \dfrac{3 + 2m}{15}$

16. Solve the inequality $-3x + 5 \leq -7$.

17. Solve the system of equations $\begin{array}{l} x + 2y = 7 \\ 3x - 3y = -6 \end{array}$.

18. What are the coordinates of the point that is 4 units to the left and 2 units below the point $(-5, 8)$?

**For Exercises 19–21, graph each equation or inequality.**

19. $y = -2x + 3$
20. $y \geq x + 5$
21. $y = 2x^2 + 4$

22. What is the slope of the line through the points $(6, 3)$ and $(4, 7)$?

23. Find the slope and $y$-intercept of the line $3x - 2y = 20$.

24. What is the perimeter of a square with area 64 in.$^2$?

25. What is the area of a square with perimeter 40 in.?

26. The lengths of the legs of a right triangle are 21 in. and 28 in. What is the length of the hypotenuse?

# Section 12.1 Exploring Algorithms

**GOAL**

**Learn how to...**
- write and use algorithms

**So you can...**
- understand real-world problems

### Application

How does a robot in a factory work? How does an automatic mail-sorter read a code and sort the mail properly? After you give it your instructions, how does a VCR really do its job? Each one of these tasks requires an algorithm that breaks the job down into smaller, more manageable parts.

### Terms to Know

**Algorithm** (p. 509)
a step-by-step method for accomplishing a goal

### Example / Illustration

To play track 8 of a CD on a CD player, you might follow these steps:

1. Open the CD player.
2. Insert the CD.
3. Press Fast Forward until "8" is displayed.
4. Press Play.

## UNDERSTANDING THE MAIN IDEAS

When you walk up a set of stairs you do not think about how to walk up the stairs, the movements are automatic. But to teach a robot to walk up a set of stairs you need an algorithm.

### Example 1

Write an algorithm for doing your laundry. Start with a pile of dirty laundry in your room. End with clean clothes put away in their proper places.

> ### Solution
>
> Here is one possible algorithm.
>
> - Put the laundry in a laundry basket.
> - Carry the laundry basket to the washing machine.
> - Open the lid of the washing machine.
> - Put in the laundry and spread it around so the items are evenly distributed.
> - Measure the laundry detergent and pour it over the laundry.
> - Close the lid of the washing machine.
> - Adjust the dials for the type of laundry and the size of the load.
> - Start the washing machine.
> - Wait for the washing machine to stop.
> - Take the wet laundry out of the washing machine and put it in the dryer.
> - Adust the dials for the type of laundry.
> - Start the dryer.
> - Wait for the dryer to stop.
> - Check the laundry for dryness. If it is dry, remove it from the dryer, fold it, and put it away. If it is not dry, set the dryer for more time and repeat the two previous steps.

1. Write an algorithm for programming a VCR to record a half-hour television show tomorrow night at 8:30 P.M. on Channel 2.
2. Write an algorithm for feeding a dog.
3. Explain how you could feed a dog using a different algorithm than the one you wrote for Exercise 2.

Sometimes when normal tools do not work, we have to think about algorithms.

> ### Example 2
>
> Suppose you need to multiply 23 × 15 but the multiplication key on your calculator is broken. Write an algorithm that you can follow to do the problem using some other calculator function keys.

**Study Guide**, ALGEBRA 1: EXPLORATIONS AND APPLICATIONS
Copyright © McDougal Littell Inc. All rights reserved.

> **■ Solution ■**
>
> Think of multiplication as repeated addition. Add 23 fifteen times. *Note:* On some calculators you can enter 23, press the addition key, and then press the equals key 14 times. (It is only 14 times because the first 23 is entered initially.)

**For Exercises 4–6, write an algorithm for solving each arithmetic problem.**

4. Divide 1245 by 234 without using the division key.

5. Multiply 71 by 9 without using the multiplication key.

6. Raise 4 to the fifth power without using the exponent key.

7. **Writing** Explain why algorithms are of special interest in this age of computers.

## Spiral Review

**Graph each inequality.** *(Section 7.5)*

8. $x < -3$

9. $y > -3x + 2$

10. $2x - 3y \geq 12$

**Simplify each expression.** *(Section 11.5)*

11. $\dfrac{4z}{3y} \cdot \dfrac{2z}{3x}$

12. $\dfrac{5y}{3x} \cdot \dfrac{6x^2}{10y}$

13. $\dfrac{4r}{2s} \div \dfrac{3s}{4r}$

14. $\dfrac{a-b}{b-a}$

15. $\dfrac{x^2 - y^2}{x+3} \cdot \dfrac{2x+6}{x+y}$

# Section 12.2 Finding Paths and Trees

**GOAL**

**Learn how to . . .**
- find the best path or tree for a graph

**So you can . . .**
- minimize the distance you travel on a trip
- connect a group of computers with the least amount of cable

## Application

You need to make a trip between several buildings at school. How can you make the trip without going over the same ground twice? You need to connect several computers with as little cable as possible. You can solve both of these problems by learning about paths and trees.

### Terms to Know

### Example / Illustration

| Terms to Know | Example / Illustration |
|---|---|
| **Graph** (p. 516)<br>a model of a map consisting of points and lines | The figure below is a graph of a city park.<br>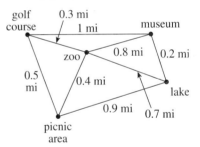<br>There are 5 vertices and 8 edges in the graph. |
| **Vertex** (p. 516)<br>a point on a graph | |
| **Edge** (p. 516)<br>a line on a graph | |
| **Path** (p. 517)<br>a collection of edges on a graph that connect one vertex to another without repeating any vertex or edge | The shortest path from the picnic area to the museum goes through the vertex labeled "lake" and is 1.1 mi long. |

Study Guide, ALGEBRA 1: EXPLORATIONS AND APPLICATIONS
Copyright © McDougal Littell Inc. All rights reserved.

## Terms to Know

| | |
|---|---|
| **Greedy algorithm** (p. 517)<br>a method for choosing a short path between the vertices of a graph (A greedy algorithm always chooses the shortest edge from the ones currently available.) | A short path connecting all five sites starts at the museum and goes to the lake, the zoo, the golf course, and then the picnic area, in that order. |
| **Tree** (p. 518)<br>a graph connecting all the vertices with a minimum of edges | 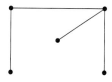 |

## UNDERSTANDING THE MAIN IDEAS

You can use graphs to represent and analyze travel situations and problems where you need to connect different machines. A graph displays all the information you need in order to determine the shortest or least expensive connections to make.

The only known way to find the *best* path in a graph is to try all the possible paths. A greedy algorithm does not guarantee the best path, but it will usually find a good path.

### Example 1

Use the greedy algorithm to find a short path between all the vertices in the graph at the right. Start at vertex $A$.

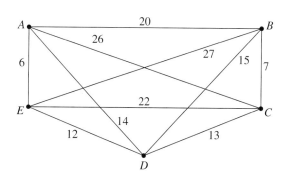

### Solution

Beginning at vertex $A$, we choose the edge representing the least distance leading to a vertex that has not been visited yet. Repeating this examination at each vertex we reach, the resulting path starts at vertex $A$ and goes through vertices $E$, $D$, and $C$ before ending at vertex $B$.

The total distance is $6 + 12 + 13 + 7$, or 38 units.

**Use the set of vertices shown at the right.**

1. Draw a path.

2. Draw a path different from the one you drew for Exercise 1.

3. Draw a tree.

4. Draw a tree different from the one you drew for Exercise 3.

**For Exercises 5–7, use the graph at right.**

5. Use a greedy algorithm to create a short path between these cities.

6. What is the shortest tree connecting vertices $A$, $B$, $D$, and $E$?

7. What is the shortest path for the graph?

8. **Writing** Explain the difference between a tree and a path. Can they be the same?

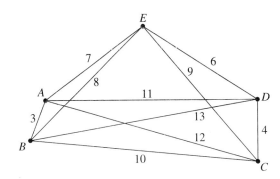

Graphs can model aspects of architecture.

### Example 2

Model the floor plan of a house shown below with a graph whose edges indicate that two rooms are directly connected. The rooms are lettered.

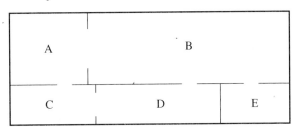

### Solution

The rooms are modeled by vertices and the doorways are modeled by edges. That is, if there is a doorway between rooms A and B, then there is an edge between vertices $A$ and $B$. The graph is shown below.

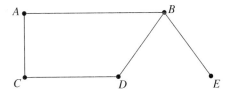

**Study Guide**, ALGEBRA 1: EXPLORATIONS AND APPLICATIONS
Copyright © McDougal Littell Inc. All rights reserved.

9. Draw a graph for the floor plan at the right that models direct access between two rooms.

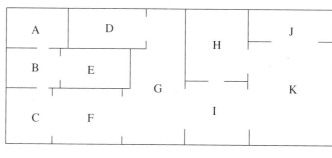

10. Can you trace *all* the edges of the graph at the right without lifting your pencil? (You can go through a vertex more than once.)

11. Draw a graph of your own that can be traced without lifting your pencil.

12. Draw a graph that cannot be traced without lifting your pencil.

## Spiral Review

**Solve each equation.** *(Section 4.4)*

13. $1.5x - 4.6 = -11.8$

14. $\frac{3}{8}x - 7 = 11$

**Solve each equation by factoring.** *(Section 10.5)*

15. $x(x + 3) = 0$

16. $x^2 - 16 = 0$

17. $x^2 - 6x + 8 = 0$

# Section 12.3 — Voting and Fair Division

**GOAL**

**Learn how to...**
- recognize fairness in elections and divisions

**So you can...**
- choose the best voting method for school elections

## Application

The election for student body president is next week. There are five candidates running, and you like two of them very much. You would be happy to see either one elected. Is there a voting method that lets you express your sentiments fully?

### Terms to Know / Example / Illustration

| Terms to Know | Example / Illustration |
|---|---|
| **One person-one vote** (p. 524)<br>a type of election in which each person can vote for one candidate | Casting a ballot in the election for the governor of your state is an example of a one person-one vote election. |
| **Approval voting** (p. 524)<br>a type of election in which each person can vote for any number of candidates (In the event of a tie, there is a tiebreaking election.) | When selecting new members to a club, the current club members can vote for each of the candidates they want to be members. This is an example of an approval voting election. |
| **Fair division** (p. 525)<br>a procedure or algorithm for dividing a group of items among a group of people | Three brothers have been given their father's baseball card collection. They have agreed to use a collector's reference book as a guide for dividing the collection evenly according to the value of the cards. |
| **Cut-and-choose** (p. 525)<br>a fair-division algorithm where one person cuts the property into two parts, then the second person gets to pick one of the two parts and the cutter gets the remaining part | Marquis and Ellis are dividing up their weekly chores. Marquis divides the list into two parts and Ellis has first choice of the parts he wants to do. |

Study Guide, ALGEBRA 1: EXPLORATIONS AND APPLICATIONS
Copyright © McDougal Littell Inc. All rights reserved.

# UNDERSTANDING THE MAIN IDEAS

In one person-one vote elections with several candidates, voters may be faced with a dilemma. They may like several candidates equally well, but they can only vote for one. In making their choice, voters may guess which of their favorite candidates has the best chance of winning, but their guess might be wrong. Candidates with similar political characteristics often split the vote and hurt each other's chances of being elected. Approval voting solves this problem by allowing voters to express their voting choices fully.

## Example 1

A mock ballot is shown below.

| Candidate | Characteristics |
|---|---|
| 1 | Popular moderate; serious; liked by almost everybody |
| 2 | Popular moderate; serious; liked by almost everybody |
| 3 | Popular moderate; serious; liked by almost everybody |
| 4 | Popular moderate; serious; liked by almost everybody |
| 5 | Joke candidate; funny; entered election on a dare |

Suppose one hundred ballots were cast in a one person-one vote election. The same voters were then given a ballot on which they could vote for as many of the candidates as they chose to support (an approval election). The results are shown below.

*One person-one vote*

| Candidate | Votes | |
|---|---|---|
| 1 | 15 | |
| 2 | 24 | |
| 3 | 18 | |
| 4 | 13 | |
| 5 | 30 | Winner |

*Approval voting*

| Candidate | Votes | |
|---|---|---|
| 1 | 62 | |
| 2 | 81 | Winner |
| 3 | 68 | |
| 4 | 60 | |
| 5 | 35 | |

Explain what may have happened in each election.

### Solution

In the one person-one vote election, the moderate, serious candidates split the moderate, serious vote. The joke candidate, who did not represent a large number of voters, won the election. In the approval voting election, each of the moderate, serious candidates received lots of votes, but the one with the greatest approval won.

1. **Writing** Explain why several candidates can appear on more than 50% of the ballots in an approval voting election.

2. **Writing** Explain why one of the popular candidates did not win the election in the one person-one vote election shown in Example 1.

3. Prepare an explanation of approval voting for a school assembly.

## *The Borda count*

A Borda count is used for many sports awards, like the Cy Young award for professional baseball pitchers. In a Borda count, voters rank their choices, and then points are assigned to their choices based on their rank. Each candidate's points are then totalled. The winner is the candidate with the most points.

### Example 2

To choose the "most valuable player" for the girls' lacrosse team, the members of the team picked their top three candidates in order of preference. Five points were awarded to the person ranked 1, 3 points to the person ranked 2, and 1 point to the person ranked 3. Six girls were mentioned on the ballots cast. The voting results are shown below.

|  | Rank | | |
|---|---|---|---|
|  | 1 | 2 | 3 |
| Sharleene | 7 | 3 | 1 |
| Keisha | 5 | 5 | 3 |
| Michele | 10 | 0 | 4 |
| Patricia | 2 | 7 | 0 |
| Dana | 12 | 1 | 1 |
| Louise | 0 | 3 | 2 |

Who is the winner?

### Solution

Assigning the points for the rankings received:

Sharleene: $7(5) + 3(3) + 1(1) = 35 + 9 + 1 = 45$

Keisha: $5(5) + 5(3) + 3(1) = 25 + 15 + 3 = 43$

Michele: $10(5) + 0(3) + 4(1) = 50 + 0 + 4 = 54$

Patricia: $2(5) + 7(3) + 0(1) = 10 + 21 + 0 = 31$

Dana: $12(5) + 1(3) + 1(1) = 60 + 3 + 1 = 64$

Louise: $0(5) + 3(3) + 2(1) = 0 + 9 + 2 = 11$

Dana has been selected as the most valuable player.

4. **Cooperative Learning** Have each person in class write down his or her three favorite music groups or performers. Do a Borda count using the point system described in Example 2. Who is the winner?

## Fair division

If two people are going to divide some property, it can be a difficult task. One solution is to sell everything and divide the proceeds equally. Another solution is to use the cut-and-choose algorithm. In this algorithm, one of the two persons divides the property into two parts and then the other person gets first choice of the two parts.

5. **Writing** Explain why the "cutter" is motivated to cut the property fairly. What happens if the cut is unfair?

6. **Mathematics Journal** What is the connection between voting methods, fair division, and algorithms?

## Spiral Review

**Find the theoretical probability that a randomly-thrown dart that hits each target hits the shaded area.** *(Section 5.6)*

7.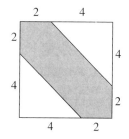
8.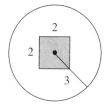
9.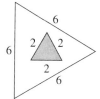

**Write each number in scientific notation.** *(Section 9.5)*

10. 0.0032
11. 9,300,000
12. 11.8 billion

# Section 12.4 Permutations

**GOAL**

**Learn how to...**
- count all the ways groups of objects can be combined

**So you can...**
- count clothing combination possibilities
- understand scheduling possibilities

## Application

A music group at school wants to produce a demo tape and send it to a radio station. There will be six songs on the tape. In how many different orders can the band arrange the songs?

### Terms to Know / Example / Illustration

| Terms to Know | Example / Illustration |
|---|---|
| **Multiplication counting principle** (p. 531) — a way to find the number of possible choices when choosing one item from each of several categories | If you have 3 pairs of sandals, 5 shirts, and 4 pairs of shorts that all go together, then there are $3 \times 5 \times 4$, or 60 different shirt-shorts-sandals outfits that can be put together. |
| **Factorial** (p. 532) — the product of all the integers from 1 to a certain number (A factorial is indicated by the symbol !.) | $5! = 5(4)(3)(2)(1) = 120$ <br> $n! = n(n-1)(n-2)(n-3)\cdots(3)(2)(1)$ |
| **Permutation** (p. 532) — an arrangement of a group of items in a definite order | There are 6 different permutations of the letters A, B, and C. The arrangements are: ABC, ACB, BAC, BCA, CAB, and CBA. |

## UNDERSTANDING THE MAIN IDEAS

The multiplication principle allows you to count all the possibilities when you have to choose one item from each of several categories. A tree diagram can also help you understand the arrangements.

*Study Guide*, ALGEBRA 1: EXPLORATIONS AND APPLICATIONS
Copyright © McDougal Littell Inc. All rights reserved.

## Multiplication Counting Principle

When you have *m* choices in one category and *n* choices in a second category, you have *m • n* possible choices. When you have *r* choices in a third category, you have a total of *m • n • r* possible choices.

### Example 1

Jeff is scheduling bands for a concert into three blocks of time. Below is a list of the bands that can play during each time block.

| 8 o'clock | 9 o'clock | 10 o'clock |
|---|---|---|
| The Wavy Tide | The Phantom Fish | The Rock Monsters |
| Peach Fuzz | Earth Farmers | Down But Not Out |
|  | Tree Hoppers | Reggae Rulers |
|  |  | The Stompers |

How many line-ups of 3 bands are possible?

### Solution

There are 2 bands to choose from for the 8 o'clock time slot, 3 bands to choose from for the 9 o'clock slot, and 4 bands to choose from for the 10 o'clock slot, so there are 2 • 3 • 4, or 24 different line-ups of 3 bands that Jeff can choose.

---

1. Darlene is setting up her class schedule for next year. She has to schedule four periods. There are six classes she can choose from for Period 1, five classes for Period 3, four classes for Period 3, and two classes for Period 4. How many different class schedules can she make?

2. Jaime has 4 pairs of shoes, 8 pairs of socks, 4 pairs of pants, and 5 shirts. How many different outfits can he put together?

3. There are two people running for president of the student council, three people running for vice-president, four people running for treasurer, and three people running for secretary. How many different groups of four officers can there be?

4. There are nine positions on a baseball team. Suppose that for the all-star team there are 10 different players competing for each position. How many possible all-star line-ups are possible for this situation? Express your answer in scientific notation and in decimal form.

5. **Open-ended Problem** Describe a situation where you would need to use the multiplication counting principle and the product 4 • 7 • 3.

## Factorials and permutations

Factorials and permutations are related. There are $n!$ permutations of $n$ different objects.

### Example 2

Suppose five students are going to line up and walk out the classroom door. In how many different orders can they line up?

### Solution

*Method 1:* Use the multiplication counting principle.

There are 5 people to choose from for the first person in line, 4 people to choose from for the second person, 3 people left for the third position, 2 people left for the fourth position, and 1 person left for the last position. So there are $5 \times 4 \times 3 \times 2 \times 1$, or 120 different line-ups.

*Method 2:* Use factorials.

There are $5! = 120$ different line-ups.

In Method 1 of Example 2, notice that the multiplication resulting from the multiplication counting principle is by definition $5!$.

**TECHNOLOGY** For Exercises 6–10, use a calculator to compute each value.

6. $6!$  
7. $10!$  
8. $20!$  
9. $25!$  
10. $30!$

11. **Writing** Explain why $3.2!$ does not make sense.

12. How many permutations are there of the letters in the name KATE? Write the ones that begin with K.

13. Fifty four students and teachers went on a field trip to a museum. In how many different ways can these 54 people line up and get on the bus?

### Spiral Review

**Solve each system of equations by adding or subtracting.** *(Section 7.3)*

14. $4x + y = 7$  
    $5x - y = 2$

15. $2x + 3y = 9$  
    $8x + 3y = 27$

**Solve each equation using the quadratic formula.** *(Section 8.5)*

16. $x^2 - 3x - 7 = 0$

17. $2x^2 + 4x - 7 = 0$

18. $-3z^2 - 5z + 11 = 0$

# Section 12.5 Combinations

**GOAL**

**Learn how to...**
- count the possible pairs of objects in a group

**So you can...**
- find the number of games in a tournament
- find the number of edges in a complete graph

## Application

Your class is going to pick 2 people out of 25 to attend a math contest. How many different pairs can there be?

### Terms to Know

| Terms to Know | Example / Illustration |
|---|---|
| **Complete graph (p. 536)** a graph in which there is an edge between every pair of vertices |  |
| **Combination (p. 537)** a selection of items chosen from a group in which order is not important | If there are 4 members of a group, then there are $\frac{4(4-1)}{2!}$, or 6 different pairs of people that can be formed. There are $\frac{4(4-1)(4-2)}{3!}$, or 4 different groups of 3 people that can be formed. |

## UNDERSTANDING THE MAIN IDEAS

### Complete graphs

A complete graph is a perfect model for many situations, such as a network of computers where every computer is connected to every other computer. Drawing complete graphs becomes increasingly difficult as the number of vertices increases.

### Example 1

Draw a complete graph that has six vertices.

**Solution**

First draw the vertices, spread out evenly. Choose one vertex and connect it to all the other vertices. Repeat the process with each of the other five vertices.

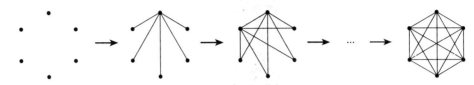

1. Draw a complete graph that has seven vertices.
2. Count the number of edges in the complete graph you drew for Exercise 1.
3. Draw a complete graph that has eight vertices.
4. Count the number of edges in the complete graph you drew for Exercise 3.

## Combinations of 2 items

The number of combinations of 2 items chosen from a group of $n$ items is given by the formula

$$_nC_2 = \frac{n(n-1)}{2!}.$$

When finding the number of permutations of several objects, the order in which they are chosen is important. When finding the number of combinations of several objects, the order is not important. For example, if two club members are chosen to be co-chairpersons of a committee, choosing Andrea and then Akeem is the same result as choosing Akeem and then Andrea.

### Example 2

Twelve computers are networked so that each computer is connected to every other computer by a cable. How many cables are there?

**Solution**

Since each cable connects two computers together and the order of the two computers is not important, we are being asked to find the number of combinations that can be made from a group of 12.

(*Solution continues on next page.*)

### Solution (continued)

Use $_nC_2 = \frac{n(n-1)}{2!}$ where $n = 12$.

$$_{12}C_2 = \frac{12(12-1)}{2!}$$
$$= \frac{12(11)}{2}$$
$$= 66$$

There are 66 cables.

---

**For Exercises 5–7, evaluate.**

5. $_{20}C_2$        6. $_{30}C_2$        7. $_{50}C_2$

8. Eight people enter a room. Each person shakes hands with every other person. How many handshakes are there? (Think of the people as vertices and the handshakes as edges.)

9. Refer to the diagram at the right. How many different acute angles can you name that have point $A$ as the vertex?

10. **Writing** Explain the connection between Exercise 9 and the number of edges in a complete graph having four vertices.

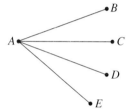

## Combinations of 3 items

The combination formula can be extended. The number of combinations of 3 items chosen from a group of $n$ items is given by the formula

$$_nC_3 = \frac{n(n-1)(n-2)}{3!}.$$

### Example 3

How many triangles can you make using six vertices?

### Solution

This is a combination problem involving choosing three items (the vertices of a triangle) from a group of six items.

*(Solution continues on next page.)*

> **■ Solution ■** *(continued)*
>
> Use $_nC_3 = \dfrac{n(n-1)(n-2)}{3!}$ where $n = 6$.
>
> $$_6C_3 = \dfrac{6(6-1)(6-2)}{3!}$$
>
> $$= \dfrac{6(5)(4)}{6}$$
>
> $$= \dfrac{120}{6}$$
>
> $$= 20$$
>
> A total of 20 triangles can be formed from six possible vertices.

**For Exercises 11 and 12, evaluate.**

**11.** $_{10}C_3$  **12.** $_{20}C_3$  **13.** $_{15}C_3$

**14.** The freshman class, with 180 students, needs a committee of 3 people to advise the school about the sports program. How many different committees could be formed?

**15.** Tom and Joyce are planning the entertainment for the class party. The video store has 500 different movies. Tom and Joyce can choose 3 movies. How many different ways can they choose the movies?

............

### Spiral Review

**The equations in Exercises 16–18 describe quantities that grow or decay exponentially. For each equation, tell whether the quantity described *grows* or *decays*. Then give the growth or decay factor for the quantity.**
*(Sections 9.2 and 9.3)*

**16.** $y = 2.4(1.12)^x$  **17.** $y = 0.7(0.95)^x$  **18.** $y = 4(3)^x$

**Decide whether the data show inverse variation. If they do, write an equation for *y* in terms of *x*.** *(Section 11.1)*

**19.**

| x | y |
|---|---|
| 2 | 10 |
| 5 | 4 |
| 0.5 | 40 |
| 10 | 2 |

**20.**

| x | y |
|---|---|
| 3 | 2 |
| 1 | 7 |
| 2 | 4 |
| 4 | 5 |

Study Guide, ALGEBRA 1: EXPLORATIONS AND APPLICATIONS
Copyright © McDougal Littell Inc. All rights reserved.

# Section 12.6 Connecting Probability and Counting

**GOAL**

**Learn how to . . .**
- use counting strategies to find probabilities

**So you can . . .**
- find the probability of winning a random drawing

## Application

Two people out of a class of 220 will be chosen as the winners of the door prizes at a dance. What are the chances that you and your best friend are the two people chosen?

## UNDERSTANDING THE MAIN IDEAS

Recall that the theoretical probability of an event is given by the ratio

$$\frac{\text{Number of favorable outcomes}}{\text{Number of possible outcomes}}.$$

The most difficult part of a probability problem is counting all the possible outcomes. It is usually easier to count the favorable outcomes. Sometimes there is just one outcome you are interested in, the outcome where you are the winner.

### Example 1

Jared's dad has a 6-sided die, an 8-sided die, and a 12-sided die. His dad said that Jared would not have to do any more work around the house this weekend if he rolled a 6 on each die when rolling all three dice. What are the chances that Jared will roll three 6's?

### ■ Solution ■

There is only one desired outcome, getting all 6's.

Use the multiplication principle to find the number of possible outcomes:

$$6 \cdot 8 \cdot 12 = 576$$

The probability that Jared will roll three 6's on the three dice is $\frac{1}{576}$.

1. What is the theoretical probability of rolling all 1's when you roll three 6-sided dice?

2. What is the theoretical probability of rolling all 1's when you roll one 8-sided die, one 10-sided die, and one 4-sided die?

3. Suppose Pablo has a 100-sided die. What is the probability that he can roll it three times and get a 2 each time?

4. A coin is like a 2-sided die on which you get heads or tails. What is the probability of tossing four coins and getting all heads?

5. Brett wondered how she would do on a 10-question true-false quiz if she flipped a coin for the answers, with heads representing "true" and tails representing "false." What are her chances of getting all ten questions correct?

### Example 2

Refer to the Application. What are the chances that you and your best friend will win the prizes?

### Solution

There is only one desired outcome, the two of you winning.

We need to know how many ways there are for 2 people to be selected from 220. This is a combination problem; use $_{220}C_2$.

$$_{220}C_2 = \frac{220 \cdot 219}{2} = 24{,}090$$

Your chances of being the two winners are $\frac{1}{24{,}090}$.

6. What are the chances that Kamala and Urmila will win the two door prizes at a party for 60 people?

7. To win a prize in a certain contest you need to choose two different numbers from 1 to 100. One pair of numbers will be the winner. What is the chance of picking the winning pair?

### Example 3

To win a prize, you must pick the three correct numbers from 1 to 50. Melissa picked the numbers 3, 9, and 49. What are her chances of winning the prize?

> **■ Solution ■**
>
> There is only one favorable outcome, the numbers 3, 9, and 49 together. To find the number of possible outcomes, we need to find the number of combinations of three numbers chosen from 50 numbers.
>
> Use $_nC_3 = \dfrac{n(n-1)(n-2)}{3!}$ where $n = 50$.
>
> $$_{50}C_3 = \dfrac{50(50-1)(50-2)}{3!}$$
>
> $$= \dfrac{50(49)(48)}{6}$$
>
> $$= 19{,}600$$
>
> Melissa's chances of winning are $\dfrac{1}{19{,}600}$.

**8.** What are the chances of picking the winning 3-number combination from the numbers 1 to 100?

**9.** One thousand people are on a luxury liner. Three people are going to win prizes at the end of the cruise. What are the chances that three friends will be the three winners?

### Spiral Review

**Simplify each expression.** *(Section 11.6)*

**10.** $(x - 5) + (8x + 3)$  **11.** $(3z - 9) - (5z - 6)$  **12.** $(3x^2 - 4x + 5) + (-4x^2 - 6x + 7)$

**Add.** *(Section 10.1)*

**13.** $\dfrac{1}{x} + \dfrac{2}{y}$  **14.** $\dfrac{x-6}{2} + \dfrac{2x-5}{3}$  **15.** $\dfrac{3}{2x} + \dfrac{5}{x}$

# Chapter 12 Review

**CHAPTER CHECK-UP**

Complete these exercises for a review of Chapter 12. If you have difficulty with a particular problem, review the indicated section.

1. Write an algorithm for washing your hair. *(Section 12.1)*

2. Explain how to use a calculator to divide 298 by 21 without using the division key. *(Section 12.1)*

3. Draw five vertices evenly spaced. Then draw a path that connects all five vertices. *(Section 12.2)*

4. Use a greedy algorithm to find a short path from vertex A to vertex B. *(Section 12.2)*

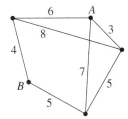

5. **Writing** Describe an election that would use approval voting. *(Section 12.3)*

6. **Writing** Explain how a Borda count works. *(Section 12.3)*

7. What does $n!$ mean? *(Section 12.4)*

8. You have 10 shirts, 3 sweaters, and 4 pairs of pants. How many different three-piece outfits can you put together? *(Section 12.4)*

9. Evaluate $_{12}C_2$. *(Section 12.5)*

10. Draw a complete graph with six vertices. *(Section 12.5)*

11. Twenty five people enter a room and each person shakes hands with every other person once. How many handshakes is this? *(Section 12.5)*

12. A 6-sided die and two 8-sided dice are rolled. What are the chances of rolling three 4's? *(Section 12.6)*

13. Pam and Sue want to win the two door prizes at a party for 26 people. What are the chances that they will be the two winners? *(Section 12.6)*

**SPIRAL REVIEW   Chapters 1–12**

Sketch the graph of each function.

1. $y = x^2 - 7x + 12$
2. $y = (x - 3)(x + 4)$
3. $y = 3(2)^x$
4. $y = \dfrac{1}{x - 1}$

**Solve each system of equations.**

**5.** $4x - 5y = 12$
$6x + 5y = 8$

**6.** $y = 3x - 2$
$2x + y = 12$

**Tonya orders softball bats for her team. Early in the season she ordered 8 Super Sluggers and 5 Power Bats. The total cost was $350. Later in the season she ordered 10 Super Sluggers and 8 Power Bats, costing a total of $490.**

**7.** Let $s$ = the number of Super Sluggers and $p$ = the number of Power Bats. Write two equations using the given information.

**8.** Solve the system of equations you wrote for Exercise 7. How much did each model of bat cost?

**Robin found the following information about the hourly wages paid in Canada.**

| Hourly wages for Canadian production workers (in U.S. dollars) | |
| --- | --- |
| 1975 | $5.98 |
| 1980 | $8.67 |
| 1985 | $10.86 |
| 1990 | $15.95 |

**9.** Graph the data. Let the horizontal axis represent the number of years since 1975.

**10.** Robin guesses that $y = 6(1.07)^x$, where $x$ = the number of years since 1975, is a good model for the growth in wages. Do you agree? Explain.

**11.** Using Robin's model, predict the wages in the year 2000.

**12. Open-ended Problem** Write a set of $x$- and $y$-values that show inverse variation.

**13.** What is the difference between a path and a tree? Can the two be the same?